Organic Companion Planting for Beginners

An Essential Guide to Growing Vegetables, Fruit, Flowers, Herbs, and More for Maximum Yield and Quality

© Copyright 2023 - All rights reserved.

The content contained within this book may not be reproduced, duplicated, or transmitted without direct written permission from the author or the publisher.

Under no circumstances will any blame or legal responsibility be held against the publisher or author for any damages, reparation, or monetary loss due to the information contained within this book, either directly or indirectly.

Legal Notice:

This book is copyright-protected. It is only for personal use. You cannot amend, distribute, sell, use, quote, or paraphrase any part of the content within this book without the consent of the author or publisher.

Disclaimer Notice:

Please note the information contained within this document is for educational and entertainment purposes only. All effort has been executed to present accurate, up-to-date, reliable, and complete information. No warranties of any kind are declared or implied. Readers acknowledge that the author is not engaging in the rendering of legal, financial, medical, or professional advice. The content within this book has been derived from various sources. Please consult a licensed professional before attempting any techniques outlined in this book.

By reading this document, the reader agrees that under no circumstances is the author responsible for any losses, direct or indirect, that are incurred as a result of the use of the information contained within this document, including, but not limited to, errors, omissions, or inaccuracies.

Table of Contents

INTRODUCTION 1
PART ONE: GETTING STARTED 2
    CHAPTER 1: THE BENEFITS OF ORGANIC COMPANION PLANTING 3
    CHAPTER 2: PLANNING YOUR GARDEN 9
    CHAPTER 3: TOOLS FOR ORGANIC COMPANION PLANTING 18
PART TWO: PLANT SELECTION AND PAIRING 25
    CHAPTER 4: COMPANION PLANTING WITH VEGETABLES 26
    CHAPTER 5: COMPANION PLANTING WITH HERBS 47
    CHAPTER 6: COMPANION PLANTING WITH FLOWERS 63
    CHAPTER 7: COMPANION PLANTING FOR PEST CONTROL 68
    CHAPTER 8: SEEDS VS. STARTERS 78
PART THREE: PLANTING, CARE AND MAINTENANCE 84
    CHAPTER 9: START WITH THE SOIL 85
    CHAPTER 10: PLANT THOSE PAIRS 96
    CHAPTER 11: WATERING AND CARING FOR YOUR PLANTS 106
    CHAPTER 12: TROUBLESHOOTING COMMON COMPANION PLANTING ISSUES 114
    CHAPTER 13: HARVESTING YOUR ORGANIC COMPANION PLANTING GARDEN 122
BONUS: ORGANIC FERTILIZER RECIPES 131
CONCLUSION 135
HERE'S ANOTHER BOOK BY DION ROSSER THAT YOU MIGHT LIKE 137
REFERENCES 138

Introduction

Companion planting is one of the oldest techniques followed by gardeners and farmers for centuries. It isn't a difficult concept; it simply means planting different plants together to improve plant health, soil structure, productivity, pest control, shade, and weed control.

Gardening isn't just about shoving plants into any old space; there is far more to it than that! You need to understand how plants work together and with their environment to create a productive, healthy garden, be it vegetables, herbs, flowers, or, ideally, a combination of all three.

This book will teach you, one step at a time, what companion planting is and how to use it to ensure your garden is the best it can be. By the end, you will have a healthy garden, which will all be done organically, with no chemicals needed.

This is an easy book to read. I've written it in plain, simple language, with complete step-by-step guides where needed and full instructions on how to do things. It's the perfect guide for beginners who don't know where to start and experienced gardeners who need a refresher or more ideas.

This is a book you can buy once and keep forever, a guide you will refer to frequently – and you should. Even the most experienced gardeners still use books for information.

So don't wait any longer. Start reading and learn how to be a fantastic gardener by understanding and using companion planting.

PART ONE: GETTING STARTED

Chapter 1: The Benefits of Organic Companion Planting

To understand the benefits of companion planting, you first need to understand what it is. It isn't a difficult concept to grasp; it's simply planting different plants together for one or more benefits, such as health, growth, pest control, etc. Those are known as good companion plants, but some don't help each other and, in some cases, can even cause problems; these are bad companions.

A Brief History

Companion planting is a centuries-old technique dating back to when agriculture first began, and evidence has been found of it in ancient civilizations worldwide. Early farmers could see the benefits of growing certain plants together to get a bigger harvest, more fertile soil, keep pests away, and have a truly balanced, organic ecosystem.

Native Americans

The "Three Sisters" companion planting technique.
Anna Juchnowicz, CC BY-SA 4.0 <https://creativecommons.org/licenses/by-sa/4.0>, via Wikimedia Commons: https://commons.wikimedia.org/wiki/File:Three_Sisters_companion_planting_technique.jpg

Perhaps the best-known companion planting example originated with the Native Americans, who developed what is now a popular technique called "three sisters." This technique involves growing beans, corn, and squash together, each plant benefiting and supporting the others. The corn provides a trellis for the beans to grow, and the beans provide nitrogen for the soil, which helps nourish the squash and corn, while the squash acts as ground cover, stopping weeds from growing and keeping the soil moist. All three plants thrived, providing Native American communities with a nutritious and sustainable food source.

Ancient Egypt

Evidence of companion planting in ancient Egypt has also been found, where farmers grew plants like garlic and onions beside their barley. Because these crops gave off a pungent odor, they kept pests away and stopped the barley crop from being damaged or destroyed. They also used peas, beans, and other legumes as cover crops to make the soil more fertile for melons, cucumbers, and other climbing plants.

Ancient China

Companion planting was a key part of ancient Chinese agricultural practices. They grew many different plants together to provide support for climbers, control pests, and improve soil health. One example from ancient China was growing beans that fix nitrogen into the soil with grain crops, like millet and rice, helping to keep the soil fertile and increase yields.

Companion planting has evolved over the years as gardeners and farmers experiment with combinations to find new combinations, and it continues to evolve today. The more scientific knowledge we gain about how plants interact, the more understanding we gain of the many benefits companion planting brings.

In the 20th century, organic gardeners made companion planting popular, along with farmers who wanted an eco-friendly, sustainable approach to agricultural practices. Today, it is one of the most widely practiced forms of organic gardening, be it in the smallest backyards to the largest farms, enhancing the health, harvest, and strength of the crops while providing a chemical-free way of controlling weeds, pests, and fertilizing the soil.

The Benefits of Companion Planting

Companion planting offers plenty of benefits, the main ones being:

- **Disease Suppressant and Pest Repellent:** Some plants give off chemicals from their roots, flowers, or leaves, which keep pests away from nearby plants and suppress certain diseases.
- **Nitrogen Fixing:** Legumes, such as beans and peas, help fix nitrogen in the soil. Rhizobium bacteria is produced by the root system, which extracts the nitrogen from the air, embedding it in the earth to fertilize it. The bacteria give some of that fertilizer to the legume plant in exchange for sugars the plant produces by photosynthesis. This is called a symbiotic relationship because the bacteria and plant benefit, and the nitrogen in the soil also helps other plants nearby.
- **Trap Cropping:** These act as pest decoys. When one plant is more attractive to a certain pest or group of pests, it can be planted near plants that the pests attack. That way, the pests will

head for the trap crop and leave the main crop alone. Trap crops are nothing more than sacrifices; if they are perennial, they will come back the next year despite the damage caused by pests, or if they are annual, they'll produce seeds or seedlings. Some trap crops are known as "dead ends" because they kill the pest once they've been trapped.

- **Masking Scents:** Many animals and insects use smell to detect food. To stop pests from eating your flowers, plant other flowers with a stronger scent to mask them. These must be planted upwind of the main plant as the pests follow scent trails on the wind.
- **Camouflage:** Other pests use the physical shape of a plant to identify it as food. Plant pest-repellent companions among your crops to mask the shape of the target crop, and plant those that attract beneficial insects as further protection.
- **Stacking:** Another benefit of companion planting is creating protective environments that protect some plants from the cold, wind, or sun and support their growth. In permaculture, plants are layered, with tall ones at the back protecting shorter or lower-placed plants from the sun. That layer of plants then provides a sheltered area for groundcover plants – that way, every plant gets the ideal conditions to grow and thrive.
- **Nurse Cropping:** Similar to stacking, nurse cropping is about planting certain plants to protect smaller, more vulnerable ones from the strong sun as they develop. They also stop the soil from eroding and prevent weed growth.
- **Biodiversity:** Another important benefit of companion planting is biodiversity. By including a good mixture of plants in your garden, you create a strong ecosystem that can survive should a pest, disease, or bad weather weaken or kill off one variety. This provides security against the entire ecosystem collapsing when one type of plant fails.
- **Maximizing Space:** Instead of having huge gaps between plants, companion plants help you maximize your space – more plants, different species, all planted together.
- **Soil Health:** Some plants help fix soil health by producing certain nutrients. We mentioned beans and peas earlier – these

add nitrogen to the soil, but other plants, such as radishes and carrots, help keep the soil loose and free.

Attracting Beneficial Insects

While some plants are used to repel pests, another way of controlling unwanted insects is to attract the beneficial ones, along with birds and arthropods, including butterflies, spiders, centipedes, and beetles.

Beneficial insects help control certain pest species, and some of the best ones to attract include:

- **Pollinators**, such as bees and certain wasps
- **Predators** that feed on destructive pests – some of the more useful are hoverflies, lacewings, ladybirds, and praying mantis
- **Arthropods** that feed on pests – predatory mites and spiders are two you should be encouraging
- **Parasites** – these attack certain pests, such as some wasp species

However, if you want a garden filled with beneficial insects, you need to attract them, and that's where certain companion plants come in. There are two types of plants these insects need:

- **Nectary** – they provide nectar as a source of food
- **Insectary** – they provide permanent homes for beneficial insects to live in and spend the colder weather

Take a cornfield, for example. Nothing but corn, as far as the eye can see. It's a fantastic place for pests that feed off corn, but it does nothing to attract and support the beneficial fauna that feeds off those pests – there are no food sources and nowhere for them to live.

Plants with small, shallow flowers are ideal for beneficial insects – daisies, calendula, carrot, parsley, dill (when allowed to flower), and plants like sweet alyssum. And for these insects to find a permanent home, you must plant perennials. We'll talk more about these later.

Good and Bad Combinations

Many crops can be used as companion plants, but getting the right combinations is imperative because they all interact differently. Experience will be your guiding factor when deciding what works best for your garden, but here are some options to get you started:

Good:
- **Beans, Corn, and Squash:** These are the 'three sisters," so named because they work in unison. The corn grows tall, and the beans can wind around the stalks. The tall corn grows quickly and also provides shade for the squash. The squash grows close to the ground and stops weeds from attacking the other two, also adding much-needed nitrogen to the soil.
- **Aromatic Herbs and Cabbages:** Insects love cabbage, so plant strong-smelling plants and flowers to mask the scent of cabbage. Mint and rosemary are perfect, but any scented herb will work.
- **Sunflowers, Cucumbers, and Radishes:** Sunflowers, just like corn, provide shade to plants and flowers below, protecting them from the harsh sun. In return, radishes and cucumbers improve soil quality.
- **Tomatoes, Basil, and Marigolds:** Basil improves the taste of tomatoes, not only after growth but when paired together in soil. Marigolds attract bees to pollinate plants while repelling pests.

Bad:

Be careful because not all combinations of plants and flowers will work.

- **Tomatoes and Potatoes:** Both are closely related; when you plant the same family of plants too close to one another, they compete for nutrients and often attract the same pests, causing an overload.
- **Brassicas and Strawberries:** Brassicas include cabbages, cauliflower, and broccoli; strawberries stop them from growing well.
- **Beans and Onions:** These require very different conditions for growing, so planting them together may result in the onions slowing down the growth of the beans
- **Cucumbers and Aromatic Herbs:** Some herbs can stunt the growth of cucumbers, and strong herbs can also change the delicate flavor of the cucumber fruit.

Keep reading, and you'll find all the information you need to be successful at companion planting.

Chapter 2: Planning Your Garden

Now you know what companion gardening is all about, it's time to start planning your garden. If you are new to this, you probably want to dive right in and get started, but there are things you need to do first.

You need to plan your garden, and for that, you need to choose a location. To do this correctly, there are a few factors to keep in mind:

1. Convenience

This is one of the most important factors when choosing a good location. If you don't have to walk too far or fight your way down to the garden, you'll have more success with growing. You'll notice watering requirements, pests, and other problems much quicker, and you'll get your harvest in on time, too.

2. Sunshine

Plants need sunlight.
https://unsplash.com/photos/dQejX2ucPBs?utm_source=unsplash&utm_medium=referral&utm_content=creditShareLink

Most plants need sunlight daily. Vegetables need at least eight hours – preferably more, to ensure they grow well and the fruit ripens. Monitor your garden to see how much sun it gets and where the partial and full shade areas are throughout the day.

Start by drawing out your garden plan and mark the sunset/sunrise hours. Head to the yard every hour and mark whether each is in full sun, shade, or partial shade. Count the hours each area is in the sun; any that don't get enough are unsuitable for vegetables.

3. Soil

Once you have decided where your garden is going, it's time to look at the soil. Healthy soil is critical for healthy plants. Ideally, you want well-drained, fertile soil. Learning about the soil in your garden is essential, so follow these steps:

Dig a Hole

Make it 12 inches by 12 inches and 6 inches deep. Put the soil you dig out onto a tarp or in a large bucket and look at it. Write down what you see:

- What colors are there? Is it loose soil or tight? Light or heavy?
- Rub a little between your fingers and write how the soil feels.

Count the Worms

Are there any worms in the soil you dug out? Have a good look – if there are at least 10, you have good fertile soil. If not, you'll need to learn how to improve the soil.

Test Drainage

Now, you want to see how well the land drains. Dig your hole 6 inches deeper – you need 12 inches for this test. Fill it with water and monitor how long it takes to drain away.

Once it has drained, repeat and monitor how long it takes to drain away again. If it's more than 8 hours, your soil drainage needs to be improved, or you could consider a raised bed or container gardening instead.

4. Water

You can use mulch and compost to help your plants become drought-resistant, but they will still need watering at some point, especially if you live in a zone that gets little rain and a lot of heat in the summer months. Seeds, in particular, need moist, warm soil for germination, and most vegetables need a steady water supply to ensure healthy growth. The ideal amount is an inch of water per plant per week.

Think about how you are going to water your garden. Is there a clean source nearby? Will you use a hose, watering cans, or drip hose? This must be considered before you get too far down the line of starting your garden.

5. Movement

The last thing to consider is how elements move through your yard. Some things to consider are:

- **Water:** How do snowmelt and rain flow through your land? Would too much wash your garden out? Does it run off or puddle and hang around, making things soggy?
- **Wind:** What direction does the wind blow in each season? Will it affect your garden, especially if it is a strong wind? Will you get weed seeds blown in from other gardens or fields nearby?
- **Equipment Access:** Can you easily access the area's necessary garden equipment? You'll need a wheelbarrow, possibly a tiller, and you might even need to have truckloads of compost unloaded and want easy access to it.

Once you have chosen your location, you should prepare it for planting.

6. Clear the Ground

Clear the ground of all grass and weeds completely, and remove any other debris and stones in the area. When planting in the fall, use layers of newspaper (up to 10) with layers of compost, potting soil, and topsoil. They can be layered or mixed. Water it well and leave it – by spring, you will have a weed-free area ready for planting.

7. Test/Improve the Soil

You can hire a professional to test your soil. However, you can invest in a kit to test your soil for smaller areas. This won't give you as much information, but it will give you a rough idea of whether your soil has sufficient nutrients or needs adjusting in some way.

Most of the time, soil in residential gardens needs a nutrient boost, especially if the topsoil has been removed for some reason. Low nutrient levels are only one thing; your soil may be poorly drained or compacted. Resolving this is easy enough; add lots of organic matter. Add a couple of inches of good compost when you till or dig a new vegetable bed. If you are working on an existing bed or don't plan on digging the soil, lay the compost over the top. Eventually, it will rot down and become organic material (humus). Earthworms will do your job for you and mix it in with the soil.

8. Prepare Your Beds

Before digging, decide what type of bed system you want - raised beds, straight rows, four-square, etc. Whatever system you go for, it's imperative to ensure the soil is loose - this will allow the plant roots to grow and pick up the nutrients and water they need. If you are planting straight into the ground, use a tiller to loosen it or get stuck in and dig it by hand. Tilling is ideal if you need to add ingredients/amendments to your soil, as the tiller will incorporate those. However, be aware that too much tilling can damage the soil structure. If your beds are small, stick to digging by hand.

You want to make this easy for yourself, so don't dig when the soil is too dry. It will be harder to get through. When the soil is too wet, it will be heavy, and it will only take more energy out of you. You want a little moisture in the dirt when you dig. Start with a garden fork to loosen the earth, and then dig it up with a spade. Turn the soil over and add in the organic matter. If you need to tread on mixed soil, lay down planks to distribute your weight.

Choose the Right Organic Fertilizer

Organic materials are great for the soil and are easy to source. Good fertilizers add nutrients gradually, working over a period of time to support plant growth. A decent product will feed your garden with macro and micronutrients, and you won't need to add chemicals.

Your plants need certain macronutrients which are found in most organic fertilizers, including the following:

- Calcium
- Magnesium
- Nitrogen
- Phosphorous
- Potassium
- Sulfur

These boost healthy growth and protect against some diseases that stunt development.

Your plants also need the following micronutrients, which are also found in organic fertilizers:
- Chlorine
- Copper
- Iron
- Manganese
- Nickel
- Zinc

These help the plants grow flowers, healthy leaves, and healthy green and yellow coloration.

This balance of macro and micronutrients cannot be found in chemical fertilizers, and chemicals don't stay in the ground long enough, so you become obliged to use them regularly – possibly causing more damage to the soil. Organic fertilizers are released slowly, working to improve water retention and soil quality over the long term.

They are also much cheaper, and you can even make your own from ingredients you already have in your home.

The Main Types of Organic Fertilizer:

Organic fertilizers can be produced from many sources, with the main ones being:
- Animal-based
- Mineral-based
- Plant-based

Animal-Based

These are typically made from animal manure and the remains left after slaughter, such as blood and bone. These are higher in nutrition than the other types and are best for leafy plants. Cow manure is the most common as it has a good balance of nutrients for all types of gardens and lawns.

Mineral-Based

These are produced from chemical processes using readily available elements from the environment. They are critical to rebalancing the composition of the soil by adding at least one macronutrient, depending on which fertilizer you use. Depending on how much you use, these can also help balance the pH level, but efficient use is required to do the most good without damaging the soil structure.

Plant-Based

As the name suggests, these are made from agricultural and plant by-products, such as molasses, green manure, cover crops, seaweed, cottonseed meal, and compost tea. They quickly break down, feed your garden plenty of nutrients, and help with soil regeneration and plant growth. They are the best choice if your garden soil is poorly drained.

How to Choose the Best

The best fertilizer is the one that matches your soil type, so you need to test your soil if you want to get it exactly right. A proper test will tell you:

- The macro and micronutrient levels in your soil
- What plants will thrive
- If you have balanced soil and, if not, what's needed to boost it.

The right organic fertilizer depends on your soil type, what you want to grow, and each plant's needs.

Garden Bed Layout Ideas

Much of how you plan your garden will come down to the available space. You will also plan around what you are planting and the maintenance needed. You can grow a garden that needs no tending, but it might not be what you want.

Rows

Rows are easy to tend to. They divide your garden neatly, and you generally run them from north to south, though you can also opt for east to west. As long as you have enough space between the rows, you can easily tend to your garden.

Tall plants like beans and corn should be planted at the north end so they don't shade other crops. Medium plants go in the middle, and smaller plants at the end. However, when you get into companion planting, this will change slightly.

Four Square

This is a simple layout, with a garden bed divided into four equal sections, each representing a separate bed. You don't need to mark these out physically if you don't want to. Each bed represents plants that require different amounts of nutrients.

Those that take a lot from the soil should be planted together and sparingly. Plants that take less can be planted together in numbers.

Rotate the crops after every season so the soil remains even over all planter boxes, and they each have the same nutrient needs. The layout is as follows:

HEAVY FEEDERS	MEDIUM FEEDERS
LIGHT FEEDERS	SOIL BUILDERS

After the first year's harvest, strip out the beds and prepare them for the next year. Each year, you will rotate the crops one square to the right, so in your second year, it will look like this:

LIGHT FEEDERS	HEAVY FEEDERS
SOIL BUILDERS	MEDIUM FEEDERS

And so on. This maintains balance in the beds.

Square Foot

Divide your grade into 4 x 4 sectors - the number of sections will depend on the size of your garden, and each will be one foot by one foot. Be sure to plant flowers that need support beside a wall or other structure. The key to this growing method is not to overcrowd each square, so be sure to check how many plants you can have in each sector.

Block

Block layouts are also known as close or wide-row planting, and they provide a much higher yield than standard row planting, with the added benefit of keeping the weeds down. The plots are similar to the square method, but the sectors can be as long as you like. This removes the need to add extra walkways, giving you more space for planting.

Using this method, you can plant a lot in a small space, but only if there is ample drainage and the plants are regularly watered. You need to tend to the plants regularly to ensure they grow and be careful to keep an eye out for pests. The rectangles can be up to 4 feet wide, but the length is only capped by your space. This makes them easy to weed and maintain. Keep walkways no more than 2 feet wide, and unless you make them out of paving slabs, add mulch to the walkways in the form of wood chips, grass clippings, or another type of organic mulch.

Make sure your plants are equally spaced in both directions. For example, a carrot patch would be spaced 3 x 3 inches. If you build a 3 x 3-foot bed, you can fit the equivalent of one 24-foot row of carrots into it – incredible space savings with a higher yield.

Vertical

Vertical gardens are ideal if you don't have much space. As the name suggests, you are planting upwards instead of horizontally. This can be done in vertical beds, baskets, or any other container that can hold soil vertically. A common method is to stack containers. This requires some work to set up but is easy to tend to once plants grow.

Containers or Raised Beds

These work well for smaller gardens or where your soil is too far gone to salvage. There are no limits to this kind of layout; the bonus of using containers is that you can move them around.

Chapter 3: Tools for Organic Companion Planting

Having the right tools is critical to getting your gardening done efficiently; it's all about making life easier so you have more time to enjoy the fruits of your labor.

Some things to consider when you purchase tools are:

- **Quality:** cheap tools won't last five minutes, so don't waste your time or money. Instead, purchase high-quality, well-made tools that will last longer. Tools often break at the joints (usually where the handle is attached), so look for one-piece tools that will last.
- **Materials:** When choosing wooden handles, hardwoods do not splinter like softwoods, so they will last longer and pose less of a hazard. Steel is great but heavy, aluminum is lighter but not as strong, and fiberglass is a good mix of the two.
- **Design:** consider ergonomic tools. For example, for those with cushioned or bent grips, consider the weight too- if they're too heavy, you won't be able to use them for too long, while tools that are too light are not likely to be strong enough.

Before shopping for tools, list everything you need and ensure you buy the best quality you can afford. Here are some ideas for useful tools:

Spade and Fork

A spade and fork are essential for marking beds and digging them.
https://unsplash.com/photos/vdD1rcsdL3E?utm_source=unsplash&utm_medium=referral&utm_content=creditShareLink

These are essential for marking your beds and digging them over. Spades can help you dig up hard soil and deep holes for trees and shrubs, while forks help you break the soil into a finer consistency. They come in all materials and sizes, so choose one that meets your requirements. Do not mistake a spade for a shovel; shovels have a flat-bottomed head, while spades are usually pointed and sharper.

Bucket

Good buckets are an excellent form of transportation. Carry your hand tools, mulch, compost, water, and even plants to where they need to go. If you can handle the weight, go for a galvanized aluminum bucket; if not, choose a sturdy plastic one.

Trug

This large, woven basket is ideal for holding your harvest and weeds, moving soil about, and much more. Some can even hold water.

Cultivator

A handle attached to a claw-like formation of metal prongs. You can use these to break up the soil, get large stones and rocks out of the ground, loosen plants for harvest, dig out weeds, and mix amendments into the soil.

You will find them in different materials, from plastic and wood to stainless steel and carbon fiber. If you want to go all out, you can opt for a two-sided tool with various tools on one side and a cultivator on the other. If you have a bad back or need to work on a larger area, you can purchase a tool with a long handle or grab an extender for your current cultivator.

Hand Fork

Excellent for loosening soil and digging over beds and containers to remove weeds, dig amendments in, and loosen the soil around plants to make for easier harvesting. Choose one made of strong plastic or metal to get the best results- some plastic ones are weak and won't last five minutes.

Footwear

The right footwear is important, so choose comfortable, durable, easy-to-clean shoes or boots. Try to have one pair dedicated purely to gardening.

Washable footwear is preferred when gardening.
https://unsplash.com/photos/tWE9W_5qTd0?utm_source=unsplash&utm_medium=referral&utm_content=creditShareLink

Non-washable shoes are not ideal as you can easily track pathogens from one garden area to another, risking plant disease. Plus, they'll get ruined fairly quickly. Clean off your footwear after every session in the garden to ensure they don't have any nasty diseases on them.

Garden Rake

Garden rakes have much firmer heads with short, strong tines, whereas a leaf rake is larger, lighter, and more flexible with long bent tines. A garden rake is dual-purpose. You can use the tined side to loosen weeds, roots, and rocks, remove dead grass, and spread the soil or soil amendments. The flat side can help you make furrows in the soil, smooth out the soil before planting, and lightly cover your seeds.

Rakes come in various sizes, so choose one that meets your requirements.

Gloves

Gloves are an important part of your gardening toolbox, but you can't wear just any old gloves – that means rubber kitchen gloves and woolly winter gloves are no good. Invest in a decent pair of proper gardening gloves. They are durable, breathable, washable, and provide a good grip. You may need a heavy-duty pair for digging, weeding, and heavy garden work and a lighter pair for sowing seeds and plants.

Wheelbarrow

A wheelbarrow can help you move equipment around the garden.
https://unsplash.com/photos/x6UXMqFw6GU?utm_source=unsplash&utm_medium=referral&utm_content=creditShareLink

Wheelbarrows make life easier because you can move equipment around the garden hassle-free. You can transport your tools, bags of compost, weeds, even buckets of water, and just about anything to help you with your gardening work.

Hoe

There are different types of hoes, and the one you choose will depend on your garden. If you focus on vegetables, you'll need a broad, strong hoe, while a perennial garden needs something lighter and thinner.

Hoes are great tools for removing weeds, especially between rows of plants, and prepping the soil in your garden beds. Traditionally, they have a flat blade with a sharp point to dig into the soil, although some have a flat bottom edge. You can also use a hoe to remove stones and rock, cover seed furrows, cut grass, make furrows, weed, and till the soil.

Hose

Lugging buckets down the garden will soon get tiresome, so invest in a good hosepipe with a multi-setting head. Make sure it is long enough to get where you need it to - you may need to join two or more together. The fittings are usually plastic, but invest in brass ones if you can get one - they last a lot longer. To save time, you can set up an irrigation system or run a soaker hose. Once these are set up, simply attach the hose and leave it to water your garden while you finish other tasks. These systems also use around 70% less water than standard hoses, and the water goes exactly where it is needed - the plant roots.

Moisture/Light/pH Meter

You can buy these separately or purchase one tool that does it all. All three are important for your garden. The pH meter tells you if the soil is right for the plants you want to grow, the light meter tells you if your plants are in too much sun or shade, and the moisture meter lets you know when it's time to water.

Tiller

These are standard tools for breaking the ground up, loosening soil, and making it easier for you to dig and plant. You can purchase manual, gas, or electric tillers, and what you buy depends on your garden. If your soil is hard and compacted and hasn't been touched in a long time, you will need a heavy-duty gas-powered tiller. However, if you have a small or medium-sized garden, you can use a smaller one to prepare the soil, remove weeds, and compost.

Pruners

Pruners are useful for cutting flowers.
https://www.pexels.com/photo/pruner-on-top-of-a-seedling-tray-6508421/

Another necessary item in your toolbox is a pair of pruners, otherwise known as *secateurs* or *shears*. These are useful for cutting flowers, pruning shrubs and plants back, and deadheading roses and other flowering plants (cutting off the old flower blooms to help signal the plant to grow new ones). Choose a good-quality pair with a sharp blade that produces a smooth, clean slice that helps the wound heal and keeps the plant healthy.

Tarpaulin

It's not necessarily a garden item, but tarps are good for many things. You can use them for covering materials and soil. Dragging plants to their new location (especially large shrubs), dragging rubble, leaves, and grass cuttings to where you want them, dragging soil or compost to the right place, storing soil you dug up while planting, lining your car trunk for when you bring plants home, and wrapping shrubs for the winter.

Trowel

An all-purpose tool, a trowel is a small spade used for small-scale cultivation. You can use a trowel for digging holes for planting, digging stones and rocks out, scooping compost into containers, doing small excavating, weeding, and transplanting. They come in all materials, different blade and handle lengths, and some have comfortable grips. Choose a full-tang trowel so it has less chance of breaking and bending. You can also purchase trowels with measurement markings on the blade, ideal for helping you gauge how deep to dig a hole.

Other Tools

You should also consider having a companion planting chart as an easy guide to see what to plant with what. Garden planning software can help you plan out your garden, while gardening apps also help you plan and identify plants and weeds and give you plenty of advice on pests and beneficial insects. A soil test kit is useful if you don't want to send your soil off for testing, while a garden journal can help you keep track of what you have planted and where, dates, varieties, notes on germination, fruiting, pruning, etc., and notes on problems with pests and other garden issues.

Lastly, purchase a selection of spray bottles for your organic fertilizers.

PART TWO: PLANT SELECTION AND PAIRING

Chapter 4: Companion Planting with Vegetables

Vegetables love to grow with companion plants, benefitting from stronger plant growth, better flavor, more yield, and fewer pests and diseases. However, it's worth noting that, while companions work, each region will differ, as will each garden, so experimentation and knowledge are key.

This chapter will list the most popular vegetables, their favorite companions, and those you should avoid.

Asparagus

Asparagus takes a few years to be established.
https://www.pexels.com/photo/flat-lay-photography-of-asparagus-351679/

Ideal for patient gardeners, asparagus takes a few years to get established, but it's well worth the wait. An established, well-cared-for asparagus bed will reap rewards in pounds of delicious vegetables.

Good Companions:
- **Basil and Parsley-** encourages vigorous growth
- **Tomatoes -** deter asparagus beetles

The tomatoes also benefit from the parsley and basil with stronger growth and more flavorful tomato fruits. Basil also deters the tomato hornworm.

Asparagus will also grow well with marigolds, comfrey, dill, and coriander, as they keep spider mites, aphids, and other pests away – do your research, though, if you grow tomatoes with asparagus. The companions must be kind to tomatoes, too.

Bad Companions:
- **Garlic and Onions -** stunt asparagus growth
- **Potatoes -** deep-rooted like asparagus and compete for space and nutrients

Beans

A good crop to grow as they add nitrogen to the soil. Some gardeners harvest half their crop and then dig the rest in to add more nutrients to the soil, but you can achieve this by harvesting all the beans, pulling the plants, chopping them up, and digging them in – do not forget the roots, as this is where the nitrogen is stored.

Good Companions:
- **Squash and Corn:** the three form the 'three sisters.' Corn will grow tall, providing shade for squash and beans and a stem for the beans to grow around. Squash keeps weeds at bay, and beans add nitrogen back into the soil.
- **Marigolds:** perfect for reeling pests. African and French marigolds exude a chemical from their roots to deter nematodes.
- **Potatoes and Catnip:** deters Mexican flea beetle.
- **Rosemary and Nasturtium:** also deter Mexican flea beetles.
- **Summer savory:** repels flea beetles and induces strong growth

and better flavor.

- **Eggplant, Radishes, and Cucumbers:** encourage growth.

Other good companions are celery, cauliflower, cabbage, broccoli, carrots, strawberries, and peas, and these also benefit from the nitrogen fixed into the soil by the beans.

Bad Companions:

- **Onion family:** this includes garlic, onions, leeks, and scallions, all inhibiting growth.
- **Kohlrabi, Basil, and Fennel** – also inhibit growth.
- **Sunflowers:** the toxins from the flowers inhibit bean growth.

Beetroot

Beets are incredibly easy to grow but require rich, fertile, well-drained soil.

Good Companions:

- **Brassicas:** the family that includes dense leafy vegetables like cabbage, broccoli, Brussels sprouts, and more. The beets provide minerals for the soil, which benefits the brassicas, and their leaves are also high in magnesium, making great compost for the brassicas.
- **Garlic:** a good deterrent against beetles, maggots, and moths. Garlic also contains an anti-fungal agent called sulfur, which protects the beets against fungal diseases. Just as it does in cooking, garlic boosts the flavor of beetroot when growing.
- **Mint:** improves beet growth, attracts predators to keep aphids away, and repels some rodents, fleas, and flea beetles. However, mint should be grown in containers as it is terribly invasive in the ground.

Bad Companions:

- **Pole Beans, Chard, and Field Mustard:** all of these will stunt the beet's growth

Broccoli

Broccoli requires regular feedings.
https://unsplash.com/photos/l55IGtwI8mI?utm_source=unsplash&utm_medium=referral&utm_content=creditShareLink

Broccoli is part of the brassica family, and it likes a lot of nutrients, which means regular feedings, especially with calcium.

Good Companions:

- **Aromatic Herbs:** rosemary, dill, potted mint, thyme, and basil all act as pest repellents.
- **Garlic:** also keeps pests away.

Bad Companions:

- **Other Brassicas:** all brassicas are heavy feeders and compete with one another for the nutrients in the soil, leaving the soil in very poor condition.
- **Asparagus, Pumpkins, Melon, and Corn:** as above, these are also heavy feeders.
- **Nightshades:** includes eggplant, peppers, and tomatoes; these will stunt its growth.
- **Strawberries and Pole Beans:** also stunt growth and compete for nutrients.

Because you can't plant much with broccoli, you tend to get too much wasted space. Maximize this by planting light feeders that won't compete for nutrients, such as marigolds, nasturtiums, bush beans, lettuce, shallots, and cucumber.

Brussels Sprouts

Also part of the brassica family, these are susceptible to many pests, making them frustrating to grow. They attract all sorts, from aphids and caterpillars to whiteflies – and many more.

Good Companions:

- **Nasturtiums:** repel some aphids, squash bugs, and whiteflies.
- **Basil:** repels mosquitos and flies and attracts beneficial insects like bees.
- **Garlic:** repels aphids and Japanese beetles and protects against blight; best planted in between Brussels sprouts for the best protection.
- **Marigolds:** repel plenty of pests.
- **Mustard:** a popular trap crop, mustard attracts many pests for Brussels sprouts. However, once attacked, the plant must be destroyed and replaced.

Bad Companions:

- **Pole Beans, Strawberries, and Tomatoes:** like all brassicas, these will stunt its growth.

Cabbage

Another brassica, cabbage is a beacon to many pests and, like all members of this family, it should be grown under fine mesh to reduce the attacks and stop butterflies from laying eggs on the leaves.

Good Companions:

- **Rosemary and Sage:** their scent repels the cabbage moth – plant these between the cabbage rows to repel pests and suppress weeds.
- **Chamomile:** improves flavor.
- **Marigolds:** repels cabbage moths, aphids, and many other pests.

- **Onions, Beets, and Celery:** repel pests and improve the flavor.

Bad Companions:

- **Tomatoes, Mustard, Grapes, and Pole Beans:** these will stunt the cabbage's growth.

Carrots

Carrots are relatively simple to grow, requiring little care except for watering and weeding.

Good Companions:

- **Tomatoes:** carrots break the soil up and aerate it, improving tomato plant growth. However, they must be planted at least 15 inches apart; otherwise, the tomatoes will stunt the carrot's growth. Tomatoes also shade the carrots and secrete solanine, a chemical that repels pests.

- **Onions, Leeks, and Garlic:** interplanting with these can help repel the dreaded carrot fly.

- **Aromatic Herbs:** chives, parsley, sage, rosemary, etc., will all repel pests.

Bad Companions:

- **Coriander and Dill:** both plants secrete chemicals into the ground that kill carrots.

- **Parsnips:** they also attract the carrot fly, so it's not a good idea to plant them together.

Plant beans in the bed the year before you plant carrots. They will fix nitrogen into the soil for the carrots to feed on. However, harvest the beans and remove the plants before planting the carrots. Otherwise, they produce too much shade and will crowd the carrots out.

Cauliflower

Yet another brassica family member that attracts a lot of pests. Rather than planting it in rows, cauliflower should be interplanted among other crops to disguise it and keep the cabbage moth away.

Good Companions:

- **Cosmos:** repels aphids and cabbage worms.

- **Nasturtiums:** good trap crop to attract aphids away from the cauliflower.
- **Fennel:** attracts parasitic wasps; these lay eggs beneath the cabbage worm's skin, thus killing them.
- **Celery and Oregano:** both repel many pests that attack brassicas.

Bad Companions:
- **Tomatoes:** because both are heavy feeders, they compete for nutrients, and neither will grow very well.
- **Strawberries:** as mentioned above, they compete for nutrients and stunt growth.
- **Other brassicas:** all heavy feeders and all attract the same pests.

Plant beans the year before to fix nitrogen into the soil for the cauliflower to feed on.

Celery

Celery is not the easiest to grow and requires a lot of water. This is why you shouldn't plant too many.

Good Companions:
- **Brassicas:** celery deters cabbage moth.
- **Leeks and Onions:** attracts insects that would attack the celery.
- **Cosmos, Snapdragons, Marigolds, Nasturtiums, and Daisies:** all repel pests and attract predators like parasitic wasps and other beneficial insects.
- **Peas and Bush Beans:** add nitrogen to the soil.
- **Tomatoes and Spinach:** provide shade to keep the soil moist.

Bad Companions:
- **Asters and Corn:** both attract diseases and harmful pests.
- **Potato and Parsnip:** heavy feeders that strip the soil of nutrients and encourage harmful pests.

Corn

Sweet corn is easy to grow but should be planted in blocks of four (not rows) because this encourages pollination.

Good Companions:

- **Beans and Peas:** they fix the nitrogen the corn needs.
- **Squash:** good ground cover to keep the moisture in and the weeds down.
- **Cucumbers:** interplanting cucumbers and corn deter raccoons.
- **Clover:** acts as mulch and a nitrogen fixer; however, be aware that clover can spread fast and will need to be controlled.

Bad Companions:

- **Tomatoes:** attract harmful pests, such as the corn earworm, which will destroy the corn crop.

Potatoes are in both camps. On one hand, the corn shades the potatoes, keeping the ground cooler and moist. However, both are heavy feeders- they will soon strip the soil and suffer unless fed regularly. Potatoes also attract a lot of pests that will eat the corn, including cutworms, potato aphids, and more.

Cucumber

Cucumber grows well in a greenhouse or poly-tunnel but is thirsty, so it won't benefit from being planted near other thirsty plants.

Good Companions:

- **Corn:** keeps raccoons away from the corn, and the cucumber will use the corn as a trellis.
- **Nasturtiums and Marigolds:** both repel harmful pests, including beetles and thrips.
- **Oregano:** repels insects.
- **Dill:** improves cucumber flavor.
- **Lettuce, Onions, and Radishes:** all repel certain insects and help improve growth and flavor.
- **Beans and Peas:** for their nitrogen-fixing abilities, especially when planted the year before.

Bad Companions:
- **Potatoes:** compete with cucumber for nutrients and water.
- **Sage:** stunts growth.
- **Tomatoes:** stunts growth and attracts harmful pests.

Eggplant

An eggplant can be grown in warmer zones.
https://unsplash.com/photos/8cqlBGw84oU

Known as *aubergine* in Europe, the eggplant is popular and has a long growing and fruiting season. It loves the sun, so it can easily be grown outside in warmer zones. In cooler climates, it must be grown in a greenhouse or poly-tunnel.

Good Companions:
- **Catnip:** deters flea beetles.
- **Hot Peppers:** secretes a chemical that prevents Fusarium diseases and root rot.
- **Sweet Peppers:** secretes fewer chemicals but has the same effect.
- **Pole Beans:** for their nitrogen, but don't let them shade the eggplants.

- **Bush Beans:** repel Colorado beetles.
- **Mexican Marigold:** also repels Colorado beetle, but it doesn't get on with beans; you'll need to plant these OR the beans, not both.
- **Thyme and French Tarragon:** repel harmful pests and garden moths.
- **Tomatoes:** similar growing requirements – don't plant too close together, though, or they will crowd each other out.

Bad Companions:
- **Geraniums:** hosts diseases that can affect the eggplant, like root rot and leaf blight.
- **Corn and Zucchini:** are both heavy feeders and will compete with the eggplant for nutrients.

Kohlrabi

Kohlrabi can be grown in cooler climates.
https://unsplash.com/photos/LYefL2BqtBY

This is a cooler weather crop and is part of the brassica family, which means it attracts a lot of different pests.

Good Companions:
- **Onions:** deters pests, including cabbage moths.
- **Lettuce:** deters earth flies.

Bad Companions:

- **Strawberries and Tomatoes:** both stunt the Kohlrabi's growth.

Leek

Part of the allium family, leeks are easy to grow and can be left in the ground until needed. You can grow leeks, garlic, and onions together, but monoculture growing can attract pests and diseases.

Good Companions:

- **Strawberries:** the strong smell from the leeks keeps pests away from the strawberries.
- **Apple Trees:** leeks also deter pests from the trees, and the apple trees improve the growth of the leeks.
- **Carrots:** this is a two-way relationship: the leeks deter carrot flies, and the carrots deter onion flies. Both crops also break the soil up, promoting good growth.
- **Parsnips:** the leeks deter pests from the parsnips.
- **Nasturtiums, Poppies, and Marigolds:** all repel pests.
- **Pepper and Tomatoes:** the leeks keep pests away from these plants and also help maximize space as they can be planted around peppers and tomatoes.
- **Beets:** similar care requirements and the leeks repel pests from the beets.
- **Celery:** both plants can grow together in trenches and have similar nutrient requirements. Leeks also keep pests away from the celery.
- **Brassicas:** do not compete for nutrients and water, and the strong smell from the leeks deters pests that attack brassicas.
- **Aromatic Herbs:** attract pollinators and deter some pests.

Bad Companions:

- **Beans and Peas:** stunt the leek's growth.
- **Asparagus:** care requirements are vastly different.

Lettuce

Easy to grow, lettuce is a popular crop among gardeners worldwide.

Good Companions:
- **Mint:** grown in pots to stop it from spreading, mint will repel slugs.
- **Onions, Carrots, and Leeks:** are all slower growing and struggle to compete with weeds; planting lettuce around these crops will smother weeds.
- **Radishes:** radishes make the lettuce taste better.
- **Cucumber:** improves flavor and provides shade for the lettuce, but don't let the cucumbers crowd the lettuce out; plant with radishes, too, as they deter cucumber beetle.
- **Strawberries:** improve the soil and bring in the beneficial insects and predators.
- **Basil:** improves growth and flavor.

Bad Companions:
- **Brassicas:** all heavy feeders and compete with lettuce for nutrients, stunting its growth.
- **Fennel:** stunts growth.
- **Parsley:** makes lettuce bolt (go to seed) very quickly.
- **Celery:** attracts the same diseases and pests as lettuce, causing damage to both crops.

Onions

Another easy plant to grow, you can start these from seed or buy seedlings.

Good Companions:
- **Brassicas:** onions repel cabbage maggots, cabbage loopers, and cabbage worms.
- **Carrots:** they help each other by keeping the onion and carrot flies away.
- **Lettuce, Strawberries, Tomatoes, and Peppers:** onions keep pests away from these crops, and they do not compete with the

onions for nutrients.
- **Parsley and Mint:** repel onion flies; do grow mint in a pot, though, as it is incredibly invasive.
- **Chamomile:** attracts beneficial insects like pollinators and repels other pests; also improves onion flavor.
- **Cucumbers, Peppers, and tomatoes:** do not compete for nutrients, and the onions keep pests away from them.

Bad Companions:
- **Beans, Peas, and Asparagus:** all require different conditions to thrive, so planting with onions will not benefit any of them.
- **Other alliums:** attract the same pests, causing an infestation.

Peas

Another popular crop – peas are good companions for lots of crops.

Good Companions:
- **Corn:** the peas can use it as a trellis.
- **Green Beans and Carrots:** require similar conditions and have no adverse effects on each other.
- **Turnips:** peas feed the soil with nitrogen for the turnips to feed on, while turnips repel pests.
- **Basil:** repels pests, especially thrips, which can decimate your peas.
- **Lettuce and Spinach:** benefit from the shade thrown by the peas and the nitrogen.
- **Cauliflower:** also benefits from nitrogen.
- **Nasturtiums:** good trap crop to keep pests away from the peas.

Bad Companions:
- **Alliums:** stunt the growth of the pea plants.

Potatoes

Potatoes are a wonderful crop to grow and, if looked after properly, can provide you with plenty of new and maincrop potatoes to keep you going through the winter.

Good Companions:
- **Chives:** attract beneficial insects and predators to attack potato pests and improve growth.
- **Cilantro:** also attracts beneficial insects and predators, like ladybirds, which feed on Colorado beetle eggs, parasitic wasps, and hoverflies.
- **Horseradish:** produces odors and chemicals that improve disease resistance.
- **Parsley and Thyme:** improve the flavor and attract beneficial insects.
- **Mint:** attracts beneficial insects and predators.

Bad Companions:
- **Nightshade family:** this includes peppers and tomatoes, which are in the same family as potatoes and compete for water and nutrients and attract the same diseases and pests.
- **Cucumbers:** they make the potatoes vulnerable to blight and compete for nutrients.
- **Sunflowers:** exude chemicals that stunt growth and seed germination.

Pumpkin

Pumpkins are easy to grow.
https://unsplash.com/photos/T9pdHqCsyoQ

Pumpkins are a favorite with most gardeners, as they are easy to grow. However, they do spread quite rapidly, so ideally, they shouldn't be planted near many plants.

Good Companions:
- **Beans and Corn:** this is the *three sisters* method mentioned earlier. The squash provides the ground cover for the other two crops, keeping moisture in the soil and crushing weeds. Pumpkins should be planted last in this method when the corn is at least 24 inches tall. Otherwise, the pumpkin will affect its growth.

Bad Companions:
- **Potatoes:** Pumpkins can cause blight in the potato crop.

Spinach

Another simple crop to grow. Simply harvest the leaves as needed, and the plant will continue growing.

Good Companions:
- **Beans or Peas:** these provide shade for the spinach and add nitrogen to the soil.
- **Tomatoes and Cucumbers:** offer shade and don't compete for nutrients with the spinach.
- **Lettuce and Strawberries:** both of these boost healthy growth in the spinach.
- **Mint:** this deters snails and slugs, the biggest pest to spinach.
- **Onion:** repels pests.
- **Carrots:** help make the soil structure better.
- **Radishes:** repels flea beetles and aphids.
- **Cilantro and Dill:** attract beneficial predators to prevent pest infestations.

Bad Companions:
- **Potatoes:** a heavy feeder, the potatoes strip the soil of nutrients and water. They also attract insects that will feast on the spinach.
- **Fennel:** stunts growth.

Squashes

The squash family includes the zucchini – these are just as easy to grow as the pumpkins.

Good Companions:

- **Corn and Beans:** most squashes create amazing ground cover and repel weeds, boosting corn and bean growth.
- **Nasturtiums:** a trap crop that attracts whiteflies, aphids, flea beetles, and other pests that might attack the squash. Plant them a distance away from the squash- too close, and the pests will hop from the flower to the squash. These flowers also make the squash fruit taste better.
- **Radishes:** deters the squash vine borer.
- **Sunflower:** provides shade.
- **Marigolds:** attract beneficial predators and deter nematodes in the soil.
- **Borage:** repels the pests, and the leaves can be mulched back into the soil to provide calcium.
- **Aromatic Herbs:** mint, dill, parsley, oregano, lemon balm, etc. These all repel plenty of pests. Make sure you grow the lemon balm and mint in pots, or they will take over the garden.

Bad Companions:

- **Melons and Pumpkins:** these compete for nutrients and water and will attract diseases and pests.
- **Beets:** this fast-growing crop can upset the squash's root system and prevent it from growing properly.
- **Fennel:** stunts the growth.
- **Potatoes:** will steal all the nutrients in the soil.

Strawberries

Easy to grow, strawberries are a favorite and produce fruit throughout the season, depending on which varieties you have. The two main types are June-bearers, which produce an earlier crop but have a short fruiting season, and ever-bearers, which produce fruit throughout a much longer season. If you have a strawberry patch, you should place netting over it to keep the birds away.

Good Companions:

- **Alliums:** onions, chives, and leeks help repel pests and help keep diseases away. Let the chives flower, and they will attract beneficial pollinators like bees.
- **Asparagus:** they share the same growing needs, but their root structures differ, so they don't interfere with one another.
- **Spinach:** both have the same growing needs and are small enough to grow in the same bed.
- **Beans and Peas:** improve nitrogen in the soil and boost growth.
- **Yarrow, Dill, Borage, Catnip, and Thyme:** all attract beneficial pollinators and predators and repel other pests while boosting plant growth and crop yield.
- **Marigolds:** repels a lot of different pests. Stick to dwarf varieties; otherwise, they will crowd out your strawberries and produce too much shade.
- **Blueberries and Cranberries:** all like the same kind of soil, and the strawberries are a kind of mulch for the other fruit bushes.

Bad Companions:

- **Mint, Okra, Tomatoes, Cucumbers, Peppers, Potatoes, and Eggplant:** all of these are prone to a disease called verticillium wilt, which can destroy your strawberries.
- **Melons and Winter Squash:** also prone to wilt, and the vines will strangle your strawberry plants.
- **Cruciferous Vegetables:** this includes cabbage, broccoli, cauliflower, chard, collard greens, and Brussels sprouts, and they can all stunt the strawberry plant's growth. Plus, they attract unwelcome pests that can decimate your strawberry harvest.

Tomatoes

Another very popular crop. Although these have been included in the vegetable section, tomatoes are, strictly speaking, fruit.

Good Companions:

- **Basil:** improves plant growth and health, makes the fruit taste better, and repels many pests, including spider mites,

hornworms, aphids, and whiteflies.
- **Borage:** improves fruit flavor and healthy plant growth while repelling the cabbage worm and hornworm.
- **Chives:** deter aphids and bring beneficial pollinators in.
- **Garlic:** deters spider mites – some people place garlic bulbs in the soil around their tomatoes to keep insects away.
- **French Marigolds:** deter slugs, nematodes, hornworms, and other nuisance pests.
- **Mint:** repels rodents, flea beetles, white cabbage moths, ants, aphids, fleas, and other pests.
- **Nasturtiums:** deter fungal infections and pests, such as aphids, squash bugs, beetles, and whiteflies.
- **Parsley:** attracts hoverflies, which feed on aphids and other pests.
- **Asparagus:** these work together; the asparagus keeps nematodes away while the tomatoes repel asparagus beetles.
- **Carrots:** break the soil up.
- **Roses:** tomatoes protect the roses from blackspot.
- **Gooseberries:** tomatoes repel insects that would attack the gooseberry bushes.

Bad Companions:
- **Brassicas:** all of these attract numerous pests that will attack the tomatoes and also stunt the growth of the tomato plants.
- **Corn:** attracts corn earworms and tomato fruit worms that will also attack the tomatoes.
- **Fennel:** stunts growth.
- **Potatoes:** tomatoes and potatoes can be affected by blight; if one gets it, the other will, too.

One other plant that can be both good and bad is dill. While it is a young plant, dill improves healthy growth in tomato plants, but it will stunt the growth when it gets older. If you want to grow dill with your tomato plants, ensure it is fully harvested while young.

Turnips

Turnips are part of the mustard family.
https://unsplash.com/photos/9c1f8Nae6j4

Also called rutabaga, turnips are a wonderful crop to grow. They are part of the mustard family and are biennial, which means they take two years to mature. The first year is spent growing the roots, leaves, and stems, while the second year produces the flowers and seeds.

Good Companions:

- **Brassicas:** the turnips are a trap crop in this case, attracting pests away from the brassicas.
- **Garlic:** turnip roots repel the borers that attack garlic, while the garlic repays this by deterring aphids, beetles, and onion flies from the turnips.
- **Beans and Peas:** add nitrogen to the soil, and because turnips are root crops and peas grow straight up, this companionship helps maximize space.
- **Nasturtiums:** attract pests away from the turnip and also attract beneficial predators and pollinators.
- **Mint and Catnip:** deters aphids and flea beetles and attracts beneficial predators and earthworms. Grow in pots as they are invasive, chop the mint leaves off regularly, and mulch them

into the soil around the turnips.

- **Thyme:** deters cabbage whiteflies and attracts beneficial pollinators and predators.

Bad Companions:

- **Potatoes:** both root vegetables; these will compete for nutrients, water, and space and will hold each other back from growing.
- **Onions:** onions are generally great companion plants, but they aren't the best to be paired with turnips because the onion bulbs grow beneath the ground, and there is a space issue. However, plant them a few feet away, and you'll benefit from them repelling pests from the turnips.

Zucchini

Also called courgettes, zucchini are incredibly easy to grow, and provided you give them the right care, you'll be rewarded with a bumper crop of these powerhouse vegetables.

Good Companions:

- **Radishes:** deter squash bugs, cucumber beetles, aphids, and many other pests. Because radishes are a fast-growing crop, you'll need to do several plantings throughout the season to reap the benefits.
- **Garlic:** repels aphids.
- **Beans and Peas:** add nitrogen to the soil.
- **Marigolds:** deters many pests and attract pollinators.
- **Nasturtiums:** the ever-popular trap crop, these will sacrifice themselves to the many predators that attack zucchini plants. Their flowers will also attract beneficial pollinators.
- **Aromatic Herbs:** including lemon balm, mint, borage, oregano, parsley, catnip, and dill; these all deter pests and attract predators and pollinators.

Bad Companions:

- **Potatoes:** stunt growth and attract pests that attack the zucchini, notably the Colorado beetle.
- **Fennel:** stunts growth.

- **Melons:** take up too much space and will crowd the zucchini out. Both plants will also be competing for nutrition.
- **Pumpkins:** compete for nutrients, and because they are the same family, there is a risk of cross-pollination, resulting in a large crop but small fruits.

As you can see, the same names crop up repeatedly as good and bad companion plants for vegetables. Most strong-smelling herbs make excellent companions because they keep the pests away, and the humble nasturtium is an excellent trap crop, constantly sacrificing itself by attracting the pests away from the main plant. I must reiterate that if you use mint, lemon balm, or catnip, you must plant them in pots or risk them taking over your gardening and smothering everything else.

In the next chapter, let's look at herbs as companions in more detail.

Chapter 5: Companion Planting with Herbs

Herbs are popular plants to grow. They make great companions for other plants, and they are also simple to grow, need very little care, and can even be used in the kitchen – fresh, dried, or frozen. Most gardeners have herbs growing in their gardens, but the fact that they make such great companion plants is a great excuse to grow even more of them.

Here are some of the best herbs you can grow and how to use them as companion plants.

Anise

Anise can help with pest control.
SABENCIA Guillermo César Ruiz, CC BY-SA 4.0 <https://creativecommons.org/licenses/by-sa/4.0>, via Wikimedia Commons: https://commons.wikimedia.org/wiki/File:Pimpinella_anisum._An%C3%ADs.jpg

Scientific Name: Pimpinella anisum

One of the more unusual herbs, anise can grow up to three feet tall and produce white lacy flowers.

Anise is excellent for pest control, repelling aphids and biting insects, and attracting beneficial insects and predators, such as predatory wasps.

Good Companions:

- **Cilantro:** they help each other geminate and produce healthy growth.
- **Brassicas:** anise uses its smell to camouflage these plants and keep pests away.

Bad Companions:

- **Basil, Beans, and Rue:** none of these grow well with anise as it stunts their growth.

Basil

Scientific Name: Ocimum basilicum

Another favorite among gardeners, basil is an excellent companion plant, grows easily in a warm sunny garden, and suits greenhouse growing. Be sure to water basil often, or it can wane and die.

Good Companions:

- **Tomatoes:** both plants improve each other's flavor.
- **Chamomile:** helps the basil grow fast and strong and increases the oil in its leaves.

Other plants you can pair with basil are:

- Chili peppers
- Asparagus
- Beetroot
- Beans
- Bell peppers
- Cabbage
- Eggplant
- Potatoes
- Oregano

- Marigolds

Bad Companions:

- **Sage and Rue:** both stunt the growth of basil.

If basil is allowed to flower, it will attract many beneficial insects to the garden. It also repels many pests, including mosquitos, hornworms, aphids, asparagus beetles, and whiteflies.

Borage

Scientific Name: Borago officinalis

Borage is a popular companion plant, mostly because it attracts pollinators and beneficial predators to the garden.

Good Companions:

- **Tomatoes and Cabbages:** repels cabbage and tomato worms, which can decimate your crops.
- **Strawberries:** helps improve the flavor of the strawberries.
- **Basil:** borage attracts pollinators and good pollinators, while basil repels insects, protecting each other. Borage also improves the flavor of basil.
- **Beans and Peas:** borage loves the extra nitrogen from the beans, returning the favor by attracting good insects and repelling the bad.
- **Cucumber, Melons, Grapes, Peppers, and Eggplant:** Borage feeds the soils with calcium and potassium, bringing the right pollinators in and repelling the pests.
- **Marigolds:** Borage grows better near marigolds; together, they are a pest-repellent powerhouse.

Bad Companions:

- **Potatoes:** if your potatoes have blight, it can kill the borage.
  - **Fennel:** at best, it will stunt the growth. At worst, it will kill the borage.

Catnip

Catnip can attract cats.
D. Gordon E. Robertson, CC BY-SA 3.0 <https://creativecommons.org/licenses/by-sa/3.0>, via Wikimedia Commons: https://commons.wikimedia.org/wiki/File:Catnip_flowers.jpg

Scientific Name: Nepeta cataria

Most people know catnip for its ability to attract cats, but it is also a type of mint. Because it attracts cats, you should cage it when you plant it near vegetables, or the cats may destroy them.

Good Companions:

- **Beans:** catnip deters Japanese beetles
- **Beets, Carrots, and Brassicas:** it also repels the flea beetles that attack these plants.
- **Lettuce:** catnip repels slugs.
- **Strawberries:** it deters many of the pests that attack strawberries.
- **Tomatoes:** benefits from the pollinators that come after the catnip flowers.
- **Squash:** any member of this family benefits because catnip can repel squash bugs.

Bad Companions:
- **Parsley:** doesn't like mint, and catnip is part of the mint family.

Chervil

Scientific Name: Anthriscus cerefolium

Chervil, also known as French parsley, is popular in France, Spain, and other Western European countries. Its leaves a taste of tarragon, parsley, anise, and licorice, and you can even eat the flowers. It grows up to two feet tall, so be careful where you plant it.

Good Companions:
- **Broccoli:** chervil improves the flavor
- **Radishes:** chervil makes the radishes crisper and hotter
- **Lettuce:** chervil enhances growth and flavor and deters ants and aphids
- **Yarrow:** enhances the essential oils in chervil
- **Alliums:** leeks, onions, and chives all help keep carrot flies, among other pests, away from the chervil

Chervil is also a trap crop and will work well when planted near vegetables that attract aphids; chervil attracts the aphids and helps protect the other plants.

Bad Companions:
- **Fennel:** fennel attracts aphids that can damage chervil and change the taste of the chervil.
- **Mint:** stunts the growth of the chervil plants.
- **Dill:** attracts pests that destroy chervil.
- **Cilantro:** similar soil requirements and will compete for nutrients.

Cilantro

Scientific Name: Coriandrum sativum

A popular kitchen herb, cilantro is also known as coriander and is a great companion to many other plants.

Good Companions:
- **Beans and Peas:** because they fix nitrogen into the soil, they help feed the cilantro. They also increase the good microbes in the soil, which helps with nutrient uptake. Peas and pole beans also provide shade.
- **Leafy Greens:** cilantro attracts good insects that benefit the greens by feeding on their pests.
- **Tall flowers:** any tall flower will work because they provide a windbreak for the cilantro and shade and act as trap crops for the pests.
- **Anise:** cilantro helps anise germinate quicker.
- **Basil and Parsley** require the same growing environment, so they are easier to care for.
- **Tomatoes:** provide shade, but don't plant them too close together as cilantro needs a lot of nitrogen, and tomatoes don't.
- **Potatoes and Eggplants:** cilantro acts as a trap crop, attracting the Colorado beetle that can destroy your plants.
- **Asparagus:** cilantro deters the asparagus beetle and improves asparagus growth.

Bad Companions:
- **Dill:** they stunt each other's growth and may cross-pollinate.
- **Fennel:** stunts growth, and they compete for nutrients.
- **Rosemary, Thyme, and Lavender** require more sun and dry soil, whereas cilantro needs less sun and moist soil.

Dill

Scientific Name: Anethum graveolens

Sadly, this herb is no longer grown as much as it used to be, but it is a wonderful companion for many plants. Be aware that it matures in just 90 days, so if you want a constant supply for the kitchen and companionship for your plants for the long term, you'll need to plant some every few weeks.

Good Companions:
- **Brassicas:** repels loopers and cabbage worms and improves plant health.
- **Alliums:** these keep aphids away from the dill.
- **Lettuce:** dill repels pests that attack lettuce.
- **Asparagus:** dill attracts beneficial insects, such as lacewings and ladybirds, to protect the asparagus.

Bad Companions:
- **Carrots:** or any other member of the umbellifer family, like parsnips as dill stunt growth, attracts carrot flies, and you risk cross-pollination.
- **Cilantro:** risk of cross-pollination as they are the same family.
- **Tomatoes:** while dill attracts parasitic wasps that feed on tomato hornworms, it can stunt the plant's growth. You can plant young dill near the tomatoes to improve growth, but pull it before it matures.

Fennel

Scientific Name: Foeniculum vulgare

Fennel is probably one of the most anti-social plants in the garden and isn't a great companion for many plants; most gardeners plant it well away from their main crops. However, it attracts many useful insects, including pollinators, parasitic wasps, ladybirds, and hoverflies. That said, some gardeners do report good results when growing fennel near other plants, so it's a case of try it and see.

Good Companions:
- **Dill:** fennel can improve growth and dill seed production, but there is a risk of cross-pollination.
- **Lemons:** fennel keeps slugs and snails away.
- **Lettuce:** fennel deters many insects, including those that attack lettuce.
- **Mint:** mint and fennel are both invasive species, so they compete for space. This results in both plants being slowed down.

- **Peas:** fennel repels pests and improves growth.

Bad Companions:

The list is too long; fennel tends to stunt growth and is an incredibly invasive plant, crowding out others.

Lemon Balm

Scientific Name: Melissa officinalis

Part of the mint family, lemon balm is an invasive species if allowed to grow unchecked – it is best planted in containers and placed where you need it.

Good Companions:

- **Melons and Squash:** Planting lemon balm in the ground can act as a natural mulch for the melons and squash, attracting plenty of beneficial insects and predators.
- **Beets:** the beets help the lemon balm grow, and the lemon balm attracts beneficial predators to protect the beets.
- **Peas:** lemon balm benefits from the nitrogen in the soil.
- **Brassicas and Tomatoes:** lemon balm attracts the beneficial predators needed to keep the brassicas and tomatoes clear of pests.
- **Radishes:** lemon balm protects the radishes from pests like snails, maggots, and aphids.
- **Carrots:** these don't compete for space, and the lemon balm protects the carrots from carrot flies and other pests.
- **Fruit trees:** lemon balm planted around the base can act as a mulch.
- **Lettuce:** lemon balm deters the pests that prey on lettuce.

Bad Companions:

- **Lavender and Rosemary:** they like different soil conditions – lemon balm likes wet soil, and lavender and rosemary like it dry; planting them together ensures one will die.
- **Fennel:** it stunts the growth of the lemon balm.

Marjoram

Scientific name: Origanum majorana

Marjoram is a wonderful herb as it attracts a lot of pollinators, making it an excellent companion for many plants.

Good Companions:

- **Squash, Zucchini, and Pumpkins:** marjoram improves the taste of all these while deterring pests.
- **Corn:** marjoram repels pests that attack corn.
- **Eggplant:** marjoram deters aphids and spider mites from destroying the eggplant fruit.
- **Onion:** marjoram improves the taste.
- **Peas:** marjoram benefits from the nitrogen in the soil and acts as a living mulch; it also attracts beneficial pollinators.
- **Potatoes:** marjoram helps keep many potato pests and diseases at bay.
- **Nettles:** if you are growing marjoram for its oil, nettles will improve production, and the nettle leaves make great liquid compost.

Bad Companions:

- **Fennel:** fennel stunts growth
- **Tomatoes:** while marjoram deters pests from the tomatoes, it requires less water than tomatoes.

Mint

Scientific Name: Mentha

Mint is an invasive plant.

https://unsplash.com/photos/boadZKqd1YM?utm_source=unsplash&utm_medium=referral&utm_content=creditShareLink

While it is incredibly invasive, mint is also an excellent herb for the garden. It comes in different varieties, including spearmint, orange, apple, chocolate, pineapple, and more.

Good Companions:

- **Kale:** mint deters aphids, cabbage moths, and other pests from the kale.
- **Brassicas:** mint attracts the predators that prey on the pests attacking the brassicas.
- **Peppers and Tomatoes:** mint repels many pests and attracts beneficial insects to the plants.
- **Eggplants:** this combination helps build a light soil structure with plenty of nutrients.
- **Roses:** mint can act as a living mulch – *but be aware it is invasive*. It also keeps the soil aerated and moist.

- **Beets:** the mint deters the pests that attack the beets beneath the ground.
- **Carrots:** mint deters aphids and carrot flies.

Bad Companions:

- **Lavender and Rosemary:** mint needs moist soil, while lavender and rosemary like it dry

Most plants will get on fine with mint so long as you grow invasive species in pots. Otherwise, they will crowd each other out, and nothing will grow properly. If you plant mint in pots, it can be placed anywhere in the garden, acting as an excellent pest repellent.

Oregano

Scientific Name: Origanum vulgare

A popular herb, oregano makes a good companion for most vegetables.

Good Companions:

- **Parsley:** stops oregano from spreading too much and improves its flavor.
- **Tarragon:** this repels the pests that would otherwise attack the oregano and provides the nutrients the oregano needs to thrive.
- **Chives:** they enhance each other's flavors. Chives also repel pests from the oregano, and oregano protects the chives from too much sun.
- **Cucumber:** shades the oregano and stops it from spreading too far, and the oregano keeps cucumber beetles away.
- **Strawberries:** the oregano keeps pests away from the strawberries, and the strawberries provide ground cover.
- **Cabbage:** the oregano keeps pests away from the cabbages, and cabbage provides some shade for the oregano.
- **Watermelons:** watermelons protect the oregano from the sun and provide a climbing place, while the oregano keeps the pests away and brings beneficial insects in.
- **Peppers:** each repels pests from the other and requires the same growing condition; oregano also improves the flavor of the peppers.

- **Beans:** oregano repels pests and enhances the growth of the bean plants, and the beans provide the oregano with nitrogen.
- **Asparagus:** oregano improves flavor and acts as a pest repellent, while asparagus shades the oregano, loosens the soil, and makes drainage better.
- **Tomatoes:** oregano repels tomato pests and is a natural fertilizer.

Bad Companions:
- **Mint:** both have different moisture requirements but are both invasive species.
- **Chives:** compete for the same nutrients; neither will thrive.
- **Basil:** has different moisture requirements; they will grow well together if grown in pots.

Parsley

Scientific Name: Petroselinum crispum

Parsley is easy to grow and has several varieties, all with unique flavors.

Good Companions:
- **Asparagus:** they each improve growth in the other, and parsley repels pests, such as the asparagus beetle.
- **Tomatoes:** parsley attracts hoverflies that attack aphids and acts as a trap crop.
- **Peppers:** parsley deters pests and improves the pepper's flavor.
- **Corn:** parsley repels pests that attack corn and attacks parasitic wasps and other beneficial predators.
- **Chives:** protect the parsley from carrot flies.
- **Basil:** parsley improves its flavor and repels some pests.
- **Beans:** parsley benefits from nitrogen and, in return, acts as a pest repellent.
- **Brassicas:** parsley deters cabbage worms from attacking your crop.
- **Roses:** parsley protects roses from aphids and many other pests.

- **Fruit trees:** parsley repels the codling moth, gypsy moth, and other pests that attack fruit trees.

Bad Companions:

- **Mint:** too invasive and will crowd out the parsley unless planted in pots, and will also affect the flavor of the parsley.
- **Carrots:** both want the same nutrients from the soil, and both attract the same pests.
- **Lettuce:** parsley hastens bolting.
- **Alliums:** can stunt parsley's growth and affect its flavor.

Rosemary

Scientific Name: Rosmarinus officinalis

Rosemary is a popular herb and is relatively easy to grow. However, if planted in the ground, it needs regular pruning to stop it from growing too leggy.

Good Companions:

- **Brassicas:** rosemary masks the smell from the brassicas, helping confuse and deter pests.
- **Beans:** feed the nitrogen into the soil for the rosemary and provide shade. In return, the rosemary deters the Mexican bean beetle and improves the bean plant's health.
- **Carrots:** rosemary keeps pests away from the carrots and helps improve growth and flavor. In return, carrots feed the soil and improve soil structure, thus improving healthy growth in the rosemary.
- **Marigolds:** keep pests away from the rosemary.
- **Strawberries:** rosemary keeps pests away from the strawberries, and both improve each other's growth. Rosemary also improves fruit flavor.
- **Peppers:** rosemary keeps the pests away and acts as ground cover, keeping the soil moist and the weeds down.
- **Onions:** they both repel pests from each other, and rosemary makes onions taste better.
- **Parsnips:** rosemary keeps the carrot flies at bay.

Bad Companions:
- **Basil:** needs more water than rosemary.
- **Mint:** both are invasive and unless planted in containers, neither will thrive.
- **Tomatoes:** these require more water than rosemary, and rosemary can inhibit tomato growth.
- **Pumpkins:** both are prone to mildew.
- **Cucumbers:** need more water than rosemary and more nitrogen, which the rosemary can't tolerate.

Sage

Scientific Name: Salvia officinalis

Sage attracts pollinators.
https://unsplash.com/photos/3tGxWVuSRBk?utm_source=unsplash&utm_medium=referral&utm_content=creditShareLink

Sage isn't grown so much these days, but it is a wonderful herb for attracting pollinators and is easy to grow.

Good Companions:
- **Brassicas:** sage is a pest repellent and helps to protect brassicas from cabbage worms, cabbage moths, and more.
- **Carrots:** sage deters carrot flies.

- **Strawberries:** sage keeps the pests away and improves strawberry flavor.
- **Tomatoes:** sage keeps the pests away and brings the beneficial insects in.

Bad Companions:
- **Alliums:** need more moisture than sage.
- **Cucumbers:** sage stunts their growth and makes the cucumbers taste bitter.
- **Rue:** inhibits sage growth.

Tarragon

Scientific Name: Artemisia dracunculus

Tarragon is not seen much in gardens today but is an excellent companion for vegetables. When you buy seeds, ensure they are not the Tagetes lucida variety; this is a tarragon substitute and not as strong.

Good Companions:
- **Chives:** these deter pests from the tarragon, and both like the same soil and moisture conditions.
- **Eggplant:** like the same moisture levels, and tarragon improves fruit flavor.
- **Cilantro:** both like the same growing conditions, and cilantro keeps spider mites away from the tarragon.
- **Garlic:** protects tarragon against spider mites, and tarragon improves garlic growth.

Bad Companions:
- **Sage, Oregano, Rosemary, Lavender:** all these prefer drier soil and won't thrive if you plant the moisture-loving tarragon with them.

Thyme

Scientific Name: Thymus vulgaris

Thyme is very simple to grow and comes in several varieties. It is a magnet for pollinators.

Good Companions:
- **Brassicas:** the thyme repels cabbage worms, beetles, slugs, and cabbage loopers. It also attracts beneficial predators, such as ladybirds.
- **Tomatoes:** thyme repels tomato hornworms and improves tomato growth.
- **Potatoes:** thyme attracts beneficial predators that keep potato pests at bay and improves the potato's flavor.
- **Eggplants:** thyme repels hornworms, aphids, beetles, spider mites, and garden moths.
- **Strawberry:** thyme repels pests and attracts beneficial pollinators.

Bad Companions:
- **Most Herbs:** need different soil and moisture conditions.

In terms of companion plants, we'll move on to flowers next before giving you a look at how to plant as a pest repellent.

Chapter 6: Companion Planting with Flowers

You now understand that companion planting is one of the best natural ways to protect your plants from pests and diseases, among other benefits. We looked at vegetables and herbs, but what about the humble flower?

Flowers are excellent for adding to a vegetable garden. They provide a wonderful pop of color and attract the best beneficial insects and predators to protect your plants. However, there are some things you need to consider when using flowers as companion plants.

- **Blooming Time:** If you want pollinators to come into your garden, you need to be sure that the flowers bloom when your flowering vegetables do; otherwise, the vegetable flowers won't be pollinated.
- **Growing Conditions:** flowers must be grown with vegetables that like the same growing conditions, i.e., soil type, water, and light. You also shouldn't grow tall flowers that shade your vegetables all day.

Flower Categories

There are three categories of flowers, each with their own defining characteristics:

1. Annuals
2. Biennials
3. Perennials

Let's look at these in more detail so you understand what to plant and when:

Annuals

Annuals take one year to complete a life cycle; this means they grow from seed and flower and produce seed in one growing season. You can also get hardy annuals; these can be planted in the fall and will sprout in the spring.

Biennials:

Bi means two, giving a clue to the fact that these types of plants and flowers complete a lifecycle every two years. The first year is purely aesthetic, and the second is when the seeds are produced.

Perennials:

Perennials are hardier than the others and will die back after the growing season, ready to regrow the following year. These are good for attracting beneficial insects and predators.

What you grow depends on your zone, your vegetable choices, and which flowers are your favorites. Don't forget that most herbs flower, too, so you get the best of both worlds. However, this chapter will focus only on actual flowers, so let's look at some of the best for companion planting.

Calendula

Also called the pot marigolds, these are different from other marigolds. Calendulas are simple to grow and flower throughout the season so long as you deadhead them. You can also harvest the seeds, dry them, and use them the following year.

Calendulas are a pest repellent, deterring asparagus beetles and tomato hornworms. They also make an excellent trap crop as they attract aphids away from other plants.

Chamomile

Chamomile is bright and attractive, bringing in bees and other beneficial insects (including predators that will rid your garden of nasty pests). And, of course, it makes a delicious tea. You can also use cold chamomile tea as a spray on your seedlings to prevent a fungal disease called "damping off."

Chamomile can be dug back into the earth at the end of the growing season to feed it with potassium, magnesium, and calcium. You can also trim it regularly and spread the cuttings at the base of any plant to act as a mulch and feed nutrients into the soil. Chamomile also helps repel mosquitoes.

Comfrey

Wilted comfrey leaves can be used to feed other plants.

Treated by many as a weed, comfrey is actually a very useful plant. It does have a very long root system. This is brittle, and should you leave even the tiniest bit in the soil, it will grow back. Plant it in containers if you don't want it to invade your garden. Bees are attracted to comfrey flowers, and the leaves can be used to make compost tea or as mulch. You can also layer wilted leaves at the bottom of a potato trench to feed the potato plants with phosphorus, nitrogen, and potassium.

Cosmos

An easy-to-grow annual, cosmos produces an excellent, colorful display. The white and orange varieties attract pollinators and green lacewings, one of the best predators that feed on aphids, thrips, and other soft-bodied insects that destroy vegetables.

Marigolds

Common in veggie gardens the world over, marigolds are excellent at attracting beneficial insects. They grow just about anywhere and come in a few varieties. The most useful is the French marigold, which helps to deter nematodes in the soil by producing a chemical from their roots. However, it can take a couple of years for this chemical to build up sufficiently, so be patient. When the season is over, chop the plant back, but leave the roots intact. Dig the foliage into the soil, and it will break down.

Because of this quality, they are good companions for most vegetables and fruits. They also deter rabbits when you plant them around your garden. Planting them with beans helps repel the Mexican bean beetle, while they can also repel squash bugs, tomato hornworms, whiteflies, and thrips – an all-around excellent companion plant.

Nasturtiums

The perfect trap crop, the brightly colored flowers on the nasturtium attract aphids and blackflies, tempting them away from valuable vegetables they would otherwise destroy. Once infested, simply remove the nasturtium stems and destroy them. They repel many types of beetles and bugs. As a bonus, the leaves and flowers are edible.

Roses

Roses are not considered good companion plants.
https://unsplash.com/photos/dv7cSiHurKM

Roses are not great companion plants because they tend to attract a lot of pests. However, that makes them good as a trap crop when planted away from your vegetables and fruits. They are especially good at attracting aphids away from grapes. You can protect the roses from some pests by planting garlic nearby and garlic chives; the latter will also flower and attract pollinators.

Sweetpeas

Another colorful crop, there are plenty of sweetpea varieties to choose from. These work well with pole beans to attract pollinators.

Let's bring this together by looking briefly at companion planting for pest control.

Chapter 7: Companion Planting for Pest Control

As you've discovered over the last few chapters, some plants are great for deterring pests by using strong aromas to mask smells from other plants or attracting beneficial insects, pollinators, and predators that feed on pests.

Pests are undoubtedly one of the biggest problems gardeners face, and while chemicals can keep them at bay, they cause serious, sometimes irreversible damage to the ecosystem in your garden. Organic is the way forward, and companion planting is the best way.

Pests are one of the largest problems gardeners face.
https://unsplash.com/photos/8oT2MA33jsk?utm_source=unsplash&utm_medium=referral&utm_content=creditShareLink

Different plants work in different ways, and while companion plants can help with pests, some do need time to build up protection to sufficient levels. For example, marigolds need a couple of years to build up natural chemical levels in the soil to deter nematodes.

It's also worth noting that companion planting won't provide a complete solution once a pest infestation has occurred. Take onions and cabbages, for example. If you plant the onions first, they will protect your cabbages from the dreaded cabbage moth. It's advisable to plant the onions last. However, if the cabbage moth has already attacked, the onions won't do much.

Some years will be worse than others for pest infestations. The insect populations ebb and flow like the tides, and some years, you won't have much trouble at all, while other years, you might just wonder why you bothered planting a garden at all! It also depends on what is growing nearby. If your neighbor has a garden full of plants that attract pests, you'll likely have many of them in your garden, too. On the other hand, if they grow a garden full of plants that attract good insects, you will benefit, too.

It's important to note that companion planting is not a complete solution. While it will deter many pests, you must still keep a close eye on your plants and take action should infestations occur. That means using organic pest controls where needed or hand-picking insects off plants.

Beneficial Insects

Some plants deter pests, but others attract the beneficial ones that feed on the pests. Provide those insects with a good environment and home to thrive in, and you can see an increase in their population above that of the pests. The chart below shows the best insects to attract and their benefits:

INSECT	BENEFITS
Parasitic Wasps	Feed on aphids, grubs, and caterpillars
Lacewing Larvae	Feed on aphids

INSECT	BENEFITS
Ground Beetles	Feed on lots of different ground pests
Hoverflies	Feed on caterpillars, leafhoppers, and many other insects
Ladybird Larvae	Feed on aphids
Robber Flies	Feed on caterpillars, leafhoppers, and many other insects
Pollinators	To pollinate your plants for a successful crop

A key aspect of using companion planting to control pests is diversity. Don't plant onions and garlic everywhere in the hope that they will keep all the nasties away. You need a good variety of plants to attract the right insects and control the bad ones for them to work effectively.

Plants must also attract these beneficial insects throughout the growing season, not just a part of it. If your flowers bloom in June but are gone a month later, they won't protect your plants for the rest of the season. Succession planting works to a degree, but diversity is the real key. This is one of the biggest mistakes gardeners make, so ensure your companion plants flower and attract these insects right through the season to provide full protection for your vegetables and fruits. If not, the good insects will go elsewhere to find food!

An important thing to remember is that beneficial insects won't feed all the time, even though food may be in plentiful supply. At times during their life cycle, they don't feed but do need somewhere to live and survive. Providing everything these insects need throughout their entire life will entice them to live in your garden and work for you, protecting your plants for longer. Fallen logs and hedgerows are good options, or you can provide insect hotels where they can overwinter safely. This also provides ground cover that locks in heat and makes it easier for early flowers to grow (also providing food and nutrients).

If you want a sustainable hedgerow, consider growing one of the fruit bushes or dwarf/patio fruit trees.

What Do Beneficial Insects Need?

To keep these beneficial insects coming in, there are certain types of plants you need to grow in your garden:

- **Ground Cover:** plants that spread across the ground, like thyme, oregano, rosemary, and sage, will provide cover for all sorts of insects, especially ground beetles. If they can hide from their predators, they can keep working for you and your garden.
- **Shade:** many insects need protected shady areas to lay their eggs.
- **Tiny Flowers:** many pollinators and beneficial predators prefer smaller flowers, like those you get on many herbs. For example, parasitic wasps are tiny and prefer clover, fennel, cilantro, dill, thyme, and so on because their flowers are tiny.
- **Composite Flowers:** other insects prefer larger flowers, like marigolds, daisies, and chamomile, including hoverflies and predatory wasps. Mint plants are good for these, too.

Deterring Pests

Herbs are among the best companion plants. Not only can you use them in your kitchen, but they deter many pests, too. These herbs also look and smell fantastic in your garden, working with the environment and each other to provide a wall of protection. Be experimental and try new things to see what works for your garden and what doesn't.

Common Pests and Companion Plants

Here's a chart showing the common garden pests and the plants that help repel them:

PESTS	DETERRENT PLANTS
Ants	Mint Catnip Wormwood Tansy

PESTS	DETERRENT PLANTS
Aphids	Chives Catnip Cilantro Eucalyptus Chrysanthemum Fennel Marigold Garlic Mustard Mint Oregano Onion Nasturtium Feverfew
Asparagus Beetle	Pot marigold Basil Parsley Tomato Tansy Nasturtium
Bean Beetle	Nasturtium Summer savory Rosemary Marigold
Black Flea Beetle	Sage
Cabbage Looper	Eucalyptus Hyssop

PESTS	DETERRENT PLANTS
	Dill
Garlic
Peppermint or spearmint
Onion
Nasturtiums
Thyme
Sage
Wormwood
Cabbage Maggot	Radishes
Marigold
Garlic
Wormwood
Sage
Cabbage Moth	Mint
Sage
Rosemary
Hyssop
Tansy
Summer savory
Thyme
Oregano
Cabbage Worm	Tomatoes
Thyme
Celery
Carrot Fly	Alliums
Rosemary
Lettuce

PESTS	DETERRENT PLANTS
	Wormwood Sage
Colorado Potato Beetle	Cilantro Marigold Onions Nasturtiums Catnip Eucalyptus Tansy
Corn Earworm	Geranium Cosmos Marigold
Cucumber Beetle	Marigold Radishes Catnip Nasturtium Tansy
Flea Beetle	Garlic Rue Wormwood Tansy Sage Mint Garlic Catnip – steep leaves in water and spray
Flies	Tansy

PESTS	DETERRENT PLANTS
	Rue Basil
Japanese Beetle	Chives Hydrangea Tansy Rue Pansy Garlic Catnip
Leafhopper	Geranium Chrysanthemum Petunias
Mexican Bean Beetle	Petunias Summer savory Rosemary Marigold
Mice	Wormwood Tansy
Mosquitos	Garlic Rosemary Geranium Basil
Moths	Rosemary Lavender Wormwood

PESTS	DETERRENT PLANTS
Peach Borer	Garlic
Nematodes	Calendula Marigolds – it takes around 1 year for chemical levels in the soil to build
Onion Fly	Garlic
Snails and Slugs	Garlic Fennel Sage Rosemary
Spider Mites	Cilantro
Squash Bugs	Mint Catnip Nasturtiums Radishes Petunias Tansy
Squash Vine Borer	Radishes
Striped Pumpkin Beetle	Nasturtiums
Ticks	Lavender Garlic
Tomato Hornworms	Calendula Borage Marigold

PESTS	DETERRENT PLANTS
	Dill
	Petunias
White Cabbage Moth	Mint
Whitefly	Marigold
	Basil
	Thyme
	Peppermint
	Oregano

When planting for pest control, make sure you choose plants that protect throughout the whole growing season. However, you will still need to monitor your plants as there is no one-size-fits-all solution.

To finish Part Two, we'll look at whether you should use seeds or starter plants.

Chapter 8: Seeds vs. Starters

Do you start with seeds or starters/seedlings from your local nursery? This is a big decision, and you need to base your decision on several factors. For the most part, you'll use a mixture of both. Some of the things you need to consider are:

- Maturing time
- Size
- Transplantability

Let's examine these features in greater detail.

Maturity Time

Different plants take different times to reach maturity. For example, tomatoes and peppers are best grown from seedlings rather than seeds, as they can take a long time to come to fruition. If your growing season is short, you won't get any joy by starting these from seed unless you start them indoors early in the year.

In contrast, spinach and lettuce don't take long to mature. Harvest can often be achieved within 30 days of planting the seeds.

Spinach plants can be harvested within 30 days of planting the seed.
https://unsplash.com/photos/4VMqrwYfmDw?utm_source=unsplash&utm_medium=referral&utm_content=creditShareLink

Knowing the time a plant takes to mature is key to knowing whether to plant seeds or not. Typically, the faster-growing plants can be started from seed, while slow-growers are best as seedlings from a nursery.

Check the instructions on the packet to discover how long they take to grow and if they need to be started indoors. Purchasing seedlings can often make more sense if the seeds take too long to grow.

Tip

Before deciding, check the maturity time. More than 65 days, you can purchase seedlings, while less than 65 days indicates you can grow from seed. However, as you will see shortly, there are exceptions to this rule.

Plant Size

Plant size is also an indicator. Typically, the bigger the plant, the longer it will take to mature and the longer it takes before you can harvest it. These are best planted as seedlings.

In contrast, smaller plants have a shorter maturity time and can be grown from seeds. Some of these, like radishes and spinach, can be sown directly into the ground, while others are best grown in pots and transplanted when they are big enough.

Transplantability

Some plants are unhappy being moved once they have grown from the seed, as their roots are somewhat fragile. For example, legumes like beans and peas are unlikely to thrive if you try to transplant them. Although they take two to three months to mature, they are best seeded directly into the ground or pre-sprouted - that means laying them on a piece of paper towel, covering them with another, and placing them all in a plastic bag. Spritz them with water to keep them damp, and once they have sprouted, you can put them in the ground.

Other plants that take a while to grow but hate being moved are zucchini, squash, and cucumbers.

Some smaller, fast-growing plants also don't take well to being moved. These include lettuce, arugula, and other small leafy greens. You should seed them directly into the soil or purchase seedlings from the nursery.

Lastly, root vegetables don't like being moved either - beets, carrots, potatoes, etc. Not only are their roots sensitive, but they also feed on nutrients in the soil, and moving them will stop that in its tracks.

Plants to Buy from the Nursery

If you are new to gardening or simply don't have the time or interest to grow from the seed, there are some plants that you should purchase from a nursery to get your growing season started quickly.

Chives

Garlic, chives, and onions will come back every year, regardless of climate, and they can be divided - more plants for your money. However, normal chives are a little tough to grow from seed but are one of the best repellents you can have.

Large Brassicas

Cauliflower, cabbage, Brussels sprouts, mustard, kale, and collards are all large plants that take a long time to mature. These are well worth purchasing from a nursery to give you a jump start on the season. However, if you have the time and space, try growing them from seed - you'll need to start them indoors early in the year. You might just find that they taste better and are healthier plants.

Nightshades

Eggplants, peppers, and tomatoes are tough to grow from seed without a long, warm growing season. All three take a long time to grow from seed, so make sure to purchase and start the seeds early in the season.

Perennial Herbs

Thyme, rosemary, oregano, tarragon, and sage all take a while to grow from seed, and many nurseries typically take cuttings from a healthy plant to grow new ones – you can do this, too once your herbs are fully grown. Provided the nursery plants are healthy, there's nothing wrong with bringing them home and planting them. You can also buy some herb plants from grocery stores.

Swiss Chard

Swiss chard is biennial, and it will last for two years. Young plants are usually readily available in nurseries, and you can enjoy them for a couple of years.

Purchasing Seeds vs. Plants

Seeds

Buying seeds allows you to control the growing environment.
https://unsplash.com/photos/Vct0oBHNmv4

Buying seeds has an advantage – one small, cheap packet usually has plenty of seeds. Not only that, but if you grow from seed, you get to control the growing environment, and you know that your plants grow in organic, nutrient-rich soils – you don't get that with nursery plants. However, make sure you purchase seeds from companies that prioritize non-GMO, organic seeds.

Plants

Plants are more expensive, but there will be occasions when it's better than buying seeds. Ensure you buy from local nurseries or organic growers rather than the big-box stores. Store plants have usually traveled a long way to get to the store and have likely been sprayed with chemicals to keep them fresh for longer. They've also likely been fed chemical fertilizers. If you try to feed them something different, they won't like it and won't grow properly. Plants like to be treated exactly the same throughout their entire growing cycle; try to change something, and they'll go into a sulk, which could lead to death.

Tips for Buying Healthy Plants

When you need to buy plants, there are a couple of things you should do to make sure you get a strong, healthy plant:

- Choose plants that have not started blooming. Get smaller plants. These can spend time growing their root system when you plant them, rather than flowers and fruit. If you can't help but purchase flowering plants, pinch off the flowers when you transplant them.
- Remove the plant and soil from the pot and look at the roots. If it spirals around the plant, it tells you it has been in the pot for a long time and is rootbound. Plants like this will struggle to establish themselves in your garden. Make sure the roots look healthy and white.
- Look for diseases and pests on the leaves, paying close attention to the underside and the stem.
- Do not buy plants that have gone leggy, tall, and narrow. They have not had enough light or have been grown in overcrowded conditions, which means they are already stressed and unlikely to thrive.

- Buy plants certified as organic. If you can't, look for labels that say they are naturally grown. Many smaller organic growers don't have the money to become approved but don't use chemicals on their plants.
- Use local nurseries. You can ask them questions about their plants and are more likely to find organic ones. Not only that, but you support a local business rather than a big-box store. These nurseries will also have plants grown in your locale, which means you know they'll be okay in your climate.

Let's move on to the real work – growing and caring for your plants.

PART THREE: PLANTING, CARE AND MAINTENANCE

Chapter 9: Start with the Soil

Healthy plants require healthy soil; it's as simple as that. If your soil is healthy, you don't need to use so much fertilizer. Instead, the soil is filled with rich organic materials, such as decaying grass clippings, leaves, and compost. It should retain moisture but have good drainage, be loose and full of the air the plants need for their roots, and be filled with minerals to help them grow. It will be populated with living organisms that help maintain its quality.

Healthy soil will help you grow healthy plants.
https://unsplash.com/photos/fjj7lVpCxRE?utm_source=unsplash&utm_medium=referral&utm_content=creditShareLink

If your plants are happy, look at the soil first. Before we dig deep – pardon the pun – into soil health, here's a quick fix to improve your soil.

Quick Fix

As a beginner, soil health can be overwhelming. This quick fix can help you with your soil before you put your plants in:

1. **Clear the Debris:** clear away rocks, stones, and other debris. If you need to remove grass, use a sharp spade to cut it into smaller, easier-to-handle pieces first.
2. **Loosen the Soil:** if your garden hasn't been dug before, use a good spade and fork to loosen the soil. You need to go 8-12 inches down to give the roots room to grow.
3. **Add Some Organic Matter:** add aged manure and compost to feed nutrients into the soil and improve drainage. This will create pockets in the soil where oxygen can enter and mix. You need 2-4 inches – no more, no less – spread over your soil and dug in. If your garden is already well dug, simply layer the compost on top and leave it – the worms will dig it in for you.

Digging Deeper

Do you know if your soil is sandy or clay? Alkaline or acid? Nutrient-rich or poor? Those are the things you need to know to be a successful gardener. It's the only way to know if you need to make changes to make things grow better, and we're going to look into all three of those in more detail.

Soil Type

There are three basic soil types: sandy, clay, and silt.

What you are looking for is an equal mix of the three. This gives good water retention and drainage with enough space for oxygen to mix. It is also light enough to make it easier to move and manipulate. But how can you tell if you have loamy soil?

It should not be sticky, even after rain. It should be damp, crumble easily, and not crust over when dry. In contrast, clay soil is sticky, retains shape when crushed, and doesn't drain well. It cracks, goes solid in the summer, and is waterlogged in fall and winter. Sandy soil is gritty, loose, and won't hold its shape. It drains too quickly, loses nutrients, and requires amending with manure and compost.

Testing Your Soil

You could opt for an official soil test, which may not be cheap. Alternatively, use a simple DIY jar test:

1. Get several Mason jars
2. Pick several sits from your garden and dig down around 6 inches – root level for many plants.
3. Take a sample of the soil and half-fill each jar, labeling the jars with each area of the garden.
4. Fill each jar with water and leave it to one side. Once the water has soaked in, pop on the life and agitate each jar for 2-4 minutes.
5. Allow the contents to settle, and measure the layer of sediment found at the bottom of the jar. This is your sand content.
6. Do the same again after letting the mixture sit for 4 minutes. Take this measurement and subtract your previous measurement. This is your silt content.
7. Allow the jar to sit for a full day, and repeat the process, subtracting your second measurement (not the second minus the first). This is your clay content.
8. Add the three, divide each measurement by the total, and multiply by 100 to get the percentages.

You want around 40% sand and silt and 20% clay.

If the percentages are off, you can add nitrogen-rich organics to decrease the sand content.

If your soil is too silty, add coarse grit or small gravel and some compost or aged manure with additional straw.

Add coarse grit, peat moss, and compost if your soil has too much clay.

Soil Nutrition

Soil tests can also tell you how fertile your soil is. Poor fertility will lead to a poor garden. You need the soil to be packed with potassium, phosphorous, and nitrogen to aid in plant growth. When purchasing fertilizer, the letters N, P, and K on the packaging denote these three nutrients.

- **Nitrogen: N** – adds the green to your plants. The leaves and stems will benefit when they have nitrogen. Older manure is usually dense in nitrogen, along with calcium-rich organics, sea life, and blood.

- **Phosphorus: P** – the catalyst. It is great for the early stages of plant growth and will ensure strong roots, which will benefit blossoms and fruit. It can be found in both quick and slow-release.
- **Potassium: K** – the defender. It will protect all parts of the plant, adding resistance and boosting the taste of vegetables and fruits.

While your plants need these three nutrients, too much can do as much harm as too little. Be sure to research how much your plant needs and add only that.

Soil pH

If the pH of the soil is off, the nutrients will not be transferred to the plants. Too low or too high, and your soil may be nutrient deficient or toxic, neither of which suits your plants.

Try to keep the pH of your soil between 6-7. However, some plants can take a range of pH while others need a specific pH level.

There are plants that thrive in acidic soil, such as blueberry bushes, but they are generally the exception to the rule. If your soil is too acidic, garden lime will fix that up. If it is too alkaline, add in some sulfur.

Be aware that this is not a quick fix. It can take upwards of a year for the amendments to have any real effect, but you only need to change your soil pH if your plant won't grow in it.

Common Soil Amendments

- **Plant Material:** most cut organics such as grass, leaves, and all softer materials (not branches, heavy roots, etc.). They need time to decompose and can greenly be composted in the fall for planting in the spring.
- **Compost:** rotting plant materials and organic scraps. Add these to your soil a few weeks before planting to add balance.
- **Leaf Mold:** lots of nutrients are packed into decomposing leaves.
- **Aged Manure:** As it ages, it develops more nutrients and loses much of its acidity. It can be smelly to store, but add it too

soon, and you risk adding disease to your soil.
- **Coconut Coir:** conditions the soil and helps it retain moisture. This is more sustainable than using peat moss.
- **Woodchips, Bark, and Sawdust:** compost these before adding them; otherwise, they will steal the nitrogen and starve the plants
- **Green Manure:** commonly called cover crops, they improve the soil. Plant in the fall, and chop and work them in in the spring. They add nutrients and soil structure.
- **Garden Lime:** raises pH
- **Sulfur:** lowers pH
- **Wood Ash:** raises pH

The last 3 should only be used if a soil test recommends them.

Adding Organic Matter

If you add organics in the fall, they have time to break down before the spring growth period. Don't be tempted to add everything at once. The material needs to start decomposing before you add more, and it needs space to do so. Take your time, and the compost will be packed with nutrients.

If you didn't do it in the fall, make sure you do so as early in spring as possible:

1. Spread at least a 2-inch layer of organic matter onto your garden; no more than 4 inches. A garden fork will help aerate the soil as you mix it. Mix the organics with the top half foot of soil, making sure of an even distribution.
2. You need to build over time, so add a little more each year. This allows the nutrients to build up.
3. Once you have your soil and compost, add a generous amount of water.
4. Don't plant in the soil immediately. Leave the soil for 2-3 weeks before you begin planting your garden.
5. Rake off any debris that might have fallen before you begin planting, and ensure the soil is even and level. Now, you can plant!

The compost adds nutrients through microorganisms, but too much of a good thing is never good. If they develop and grow too quickly, they can deplete the nutrients instead of adding to them. This will mess up the pH level of your soil. It needs to comprise about a quarter of your soil mixture and should be mixed into your soil thoroughly.

Using Raised Beds

You could use raised beds if your soil cannot be amended quickly enough. This way, you control the soil and its nutrient levels. Do NOT walk on the soil in your beds; it will compact quickly and go hard. Keep your beds no more than 4 feet wide, or place a pathway down the center if you want wider ones.

Raised beds limit frost and ice in colder areas, and you can plant a few weeks earlier without worrying about damaging the seed pods. You can cover them with a non-porous dark material to keep weeds down and warm up the soil to start growing in them earlier.

Using Cover Crops to Improve Soil

Soil fertility is important, as you know by now, and there are some underlying principles to help you maintain healthy soil:

- Keep the soil covered as much as you can
- Do not disturb the soil unless necessary
- Keep roots growing all year around to feed the soil
- Diversify what you plant

That's where cover cropping comes into its own. Otherwise known as green manure, cover crops help you do all that and are a cost-effective way of maintaining rich, fertile garden beds and improving quality and yield.

If you want to go down the route of adding amendments every season, go right ahead. But cover cropping is a great way to achieve the same results without the heavy lifting. Simply dig your beds in the fall, scatter the seeds over the top, cover them with a thin layer of soil, and water them. Keep them moist until they germinate, and let them grow.

Once a cover crop is established, it will keep the weeds down and stop virtually all topsoil erosion caused by the wind. When the crop is ready, simply chop it, drop it onto the garden, and leave it to die a little before digging it in - including the roots. As it decomposes in the soil, it will feed it with nutrients and nitrogen, improving microbial activity and soil quality. As these crops are planted when your garden would otherwise be empty, they do all the work for you while you can sit back and take a break from the hard work.

The soil is full of living microorganisms, and cover crops feed these. However, it's a symbiotic relationship. Those microorganisms feed the plants with nutrients like phosphorus and nitrogen.

Cover crops with long taproots keep soil compaction to a minimum, thus aiding plant growth. Their roots dig deep down into the soil and aerate it, encouraging moisture to go deeper into it and reducing the runoff risk.

Which cover crops you plant depends on what you want. If your soil is clay, compacted, sandy with little fertility, suffers from soil erosion, or has insufficient organic matter, it will benefit from a cove crop, so first, work out what you need to fix.

Best Cover Crops

Cover crops are usually perennials with a short lifecycle, annuals, or biennials. They all have advantages and disadvantages, and these are some of the best ones to use, with tips on how to get rid of them.

Winter Rye

Scientific Name: Secale cereal

An annual grain, winter rye adapts to all types of soil, even soil with little fertility, sandy, and acidic soils. It is a crucial cover crop for the winter as it suppresses weeds, and its roots exude a chemical that stops weeds from germinating. It is planted as the last crop in the season because it needs cool soil to germinate, typically when the temperatures are down to 35°F. Typically, this means planting it in the fall, about the time of the first frost, any earlier, and it can turn aggressive.

How to Remove It

In the early spring, move the rye and leave it on the ground or till it is straight into the soil. A warning, though- it doesn't die in the winter, so you need to time your removal to stop it from going to seed. If you are mowing and leaving it, leave it a few weeks to decompose before you start planning in the garden.

Field Peas

Scientific Name: Pisum sativum

Peas like cool weather. Plant when the hot summer weather starts or when it has passed. Peas put nitrogen into the soil to feed future crops. Rhizobia are bacteria in the soil that convert atmospheric nitrogen into something plants can use, eventually attaching themselves to pea roots. When the peas have died back, the nitrogen stays in the soil. To plant these, bury each pea up to 2 inches deep in the soil.

How to Remove It

If you plant in the spring, the peas will die by the late spring and can stay where they are to decompose, or you can till them back into the soil before you replant. If you plant in the fall, their stems die in winter and will be fully decomposed before the spring growing season starts.

Oats

Scientific Name: Avena sativa

Oats are cool-season grass that germinates energetically, establishing itself very quickly, especially spring-grown oats. Their profuse top growth and stringy roots improve soil structure when you till them in. They scavenge phosphorus from the soil, and plantlets hoover up extra nutrients, which restore soil fertilization levels. You can plant these in the fall or spring.

How to Remove It:

When spring oats have been growing for 6 to 10 weeks, you can mow them and leave them to decompose or till them in, preferably while they still have green seed heads. Fall oats are planted from the third week of September, allowing them to be established before early winter and soil erosion hits. If you are in a cold zone, they will not survive, so you can easily till them into the ground in the spring.

Crimson Clover

Scientific Name: Trifolium incarnatum

Crimson clover is a beautiful plant that can be grown anytime throughout the growing season. It has a simple taproot that mines the soil and builds up nitrogen. It is great at suppressing weeds and controlling erosion, and if left to flower, it will attract pollinators.

How to Remove it

Once it starts to bud, mow it down and till it in. You can leave it until it flowers, but you must be quick, or it will self-seed and grow everywhere. If you plant it in the winter, it will die back naturally in cold winters.

Ryegrass

Scientific Name: Lolium multiflorum

Ryegrass is readily available in most nurseries for spring or fall planting. A packet of seed will usually be a combination of annual and perennial grass. When you plant annual ryegrass in the fall, it will die back in the winter, and you can till it into the soil in spring. The perennial seed is more difficult as its root system is long. Given the right moisture levels, it is prolific in the cool seasons, so plant in late summer through early fall. Any later than that, and it won't establish properly, especially if a frost hits.

How to Remove It

Mow it down in early spring. It may need more than one mowing if you have the combination seed, and you may need to smother it if you want to kill it off completely. Leave it to decompose for a few weeks before you plant in the soil.

Oilseed Radish

Scientific name: Raphanus sativus

This is the best cover crop for compacted soil. It breaks soil up easily with its long taproot and improves soil drainage significantly. It looks more like a daikon radish than your typical spring variety. If you plant it midsummer, it will grow big enough to create pockets in the soil, allowing better air and water filtration and making it easier for seedlings to root themselves in the sprint.

How to Remove It

It will die off once temperatures drop to 20°F or lower and will be fully decomposed by the spring.

Using Organic Fertilizer

Chemical or synthetic fertilizers are cheap and readily available, so why would you want to use the more expensive organic versions on your soil? There are several reasons why. It's not all about the quick result.

Synthetics might work right now, but organic fertilizers keep the soil healthy long-term.

They Work Slowly

Before an organic fertilizer can work, it must be broken down by the soil, ensuring the plants and soil get the right nutrition when needed. Synthetic fertilizers usually result in overfeeding, can burn plants, and do not benefit the soil.

They Improve the Soil

Organic fertilizers and materials improve the texture of the soil, helping it retain moisture and significantly increasing the activity of microorganisms. Not only do they help plants, but they help the soil, too. Synthetic fertilizers strip nutrients from the soil, providing a very poor harvest.

They Are Safe

Not to eat or drink, obviously, as most of them would taste disgusting, even though they are natural. However, they are safe for your garden, the environment, your kids, and the family pets. Conversely, synthetics use fossil fuels to produce them, and runoff often pollutes water sources.

They Are Easy to Use

They couldn't be easier; simply mix, spray, or add to the soil. They benefit your garden in many ways and are as convenient as synthetic fertilizers.

Mixing Organic Fertilizer with Soil

You need to consider a few things before you do it. The fertilizer cannot be added as is and must be watered down first. This ensures the fertilizer is evenly distributed.

You also want to add it slowly. Add a little and mix it thoroughly, starting with a ratio of 1:10 fertilizer to soil. Once the right amount has been added, you can water the soil.

First, decide what you want to use – loads of organic fertilizers are available, or you can make your own. The type of fertilizer will depend on your garden and growing needs. Make sure you follow the package directions carefully and add the fertilizer in the right amounts – do not add too much, as it can cause harm to the soil and plants.

Mix it in, using a garden fork or space to distribute it through the soil evenly.

Lastly, plant your plants and ensure you give them enough room to grow.

In the next chapter, we'll get down to the nitty-gritty of planting your plants and their companions.

Chapter 10: Plant Those Pairs

You know what you want to plant, and you've got your soil just right, so now it's time to get your hands dirty. This chapter will start by looking at planting seeds vs. seedlings.

Seeds

Growing plants from seed might seem daunting to some beginners, but it isn't that difficult. It is far more rewarding than buying plants. Once you get into it, it's fun. This section will give you the confidence to start growing your plants from seed, be they vegetables, herbs, or flowers.

So, why would you grow from seed when you can head to the nursery and pick up everything you need?

As mentioned earlier, some plants are better grown from seeds, particularly those that don't take long to mature. That said, it really is down to you how you grow your garden, but it's a guarantee that once you start growing certain plants from seed, you won't look back.

Don't pressure yourself. If things don't go right, you can always buy the plants from the nursery and have another go at growing from seed. However, to push you in the right direction, here are some of the primary benefits:

- **It's Cheaper:** buying a pack of seeds is far cheaper than buying the plants, and you get more for your money. However, keep in mind that seeds expire, so check the dates on the packets.

- **More Choice:** there will always be more choices with seeds as opposed to plants, so you get more options for your garden and way more varieties of each vegetable, herb, or fruit, too.
- **You Know What You Are Growing:** when you grow from seed, you control the growing environment and the fertilizers, pesticides, and fungicides you use. With any luck, you'll be doing it all organically. You have no idea what chemicals, if any, have been used with nursery plants.
- **You Can Start Earlier:** especially if you live in a colder climate, you get to start your seeds indoors earlier and have the pleasure of watching them grow into healthy seedlings ready for the spring.
- **Pride:** you grew those plants, so you have every right to be proud of yourself for your achievements.
- **Plenty for Everyone:** whenever you grow from seed, you always sow a few extra, just in case some don't make it. That means you'll likely have extras; share them with your friends and family or sell them to make a bit of cash for next year's seeds.

Seed Starting 101

Most beginners struggle because the technical stuff confuses them, or they try to be clever and do things in a complicated way. Growing from seed isn't that difficult, but let's clear up some of the basics for you right now!

Technical Terms:

Yes, gardening does come with a few technical terms, but they aren't difficult to understand; you'll soon be using these words like a pro! These are some of the most important ones:

- **Sowing:** nothing more than planting your seeds
- **Germination:** when your seeds begin to form seedlings
- **Scarification:** scratching a head seed's outer coating to speed the germination process
- **Stratification:** simulating cold-weather conditions for those seeds that need to be dormant in the cold before they can germinate

Seed Starting Techniques

Success at growing from seed depends on you doing it the right way, and you can use two main techniques: indoor and direct sowing.

- **Indoor Sowing:** your seeds are sown into containers and kept indoors for several weeks before transplanting the seedlings into the garden. This way, you can start your crops off much earlier than outside. This method is ideal if you want to start slow-growing crops like tomatoes and peppers from seed.
- **Direct Sowing:** with this technique, the seeds are sown directly into the soil where you will grow them; no special equipment is needed, and no need to transplant seedlings.

Equipment:

Newbies often balk at growing from seed because they think setting themselves up with the equipment will be expensive. Here's the truth: you don't need a lot.

- **Seeds:** don't go overboard; you know what vegetables you want to grow, and you've worked out your companion plants. When you've worked out which ones you will grow from seed and which you will buy as seedlings, you know what seeds to buy.
- **Soil:** you can't use the soil from your garden for this; you'll need to buy some potting soil from your nursery. This contains the right mix of nutrients required for your seeds to germinate and grow.
- **Water:** don't use tap water if you can help it; it has too much chlorine. If that's your only option, place some into a jug and leave it for 24 hours at room temperature to dissipate the chlorine. Use clean rainwater or snowmelt if possible, and bring it to room temperature before using it.
- **Seed Trays:** these are for placing your pots of seeds in.
- **Pots:** for starting seeds, you can use 3-inch pots or root trainers (for certain seeds).

Different Types of Seed

We don't mean vegetable or herb types here. Different seed types grow in different ways, but when we break it down, there are two main categories: warm weather and cold hardy:

- **Warm Weather Seeds:** these will only germinate and grow in a warm environment. Too cold, and they'll do nothing, and even if they do, the seedlings won't survive. These are usually the best ones to start growing indoors and include peppers, tomatoes, eggplant, broccoli, basil, cosmos, zinnia, and marigolds, among many others.
- **Cold Hardy Seeds:** these seeds like the temperature cooler. If it's too warm, they won't germinate, or the seedlings will likely die. These are typically directly sown into the ground outdoors and include spinach, lettuce, radish, beans, beat, carrots, peas, sunflowers, and petunias.

Preparing to Grow Your Seeds

Before you start growing, you need to be prepared. Jump straight into it without following these next couple of steps, and you may not be successful:

- **Read the Packet:** this might sound daft, but you'd be surprised how many gardeners don't read the seed packets and wonder why they fail. Each packet tells you the growing requirements for the specific seed, including when to plant, whether to start them inside or outside, when to expect harvest, and so on.
- **Get Ready:** get all of your supplies ready before you start; this includes seed trays, pots, soil, etc. If your trays are older, you must clean and disinfect them before you use them in case they carry traces of disease or tiny pest eggs.

How to Plant Your Seeds

It doesn't matter if you are sowing in pots indoors or straight into the ground; the process is pretty much the same for both:

Step One: Prepare the Soil

If you are directly sowing into the ground, loosen off the top three or four inches of soil. Dig in some compost or worm castings and the organic fertilizer of your choice. If you are starting indoors, have a bag of high-quality seed compost.

Step Two: Work Out Your Spacing

This depends on what you are growing, and the spacing requirements will be written on the seed packet.

Step Three: Get Sowing

Again, a lot depends on the plant. Some seeds should be buried deeper in the soil than others. Make your hole and drop the seed in, or put it on top of the soil and press it down. The latter won't work with carrots and other fine, tiny seeds; these can just be sprinkled over the soil.

Step Four: Cover Them Up

Cover your seeds with soil and pat it down gently.

Step Five: Water Them

Spray water over indoor seeds. You want the soil moist but not washed out. Set your hose to a fine spray for outdoor sowings and spray it lightly over; do not disturb the seeds.

Tracking Your Plants

If you purchased a garden journal, use it now. If not, start a spreadsheet on your computer or just grab a notebook and pen. Write down the following:

- The seeds you just planted
- The planting date
- The gemination date
- Success rate – how many sprouted successfully

Make notes on which techniques you used and, as you track from start to finish, note what worked and what didn't, what you could do better, issues you faced, and so on.

Seedlings

Whether you grow your seedlings from seed or purchase them from a nursery or local grower, the process of handling and planting them is the same. You will need to transplant them from their current growing environment into another one.

What Is Transplanting?

Seedlings need to be transplanted into soil.

Transplanting refers to moving a plant from one environment to another, in this case, from one pot to another, whether purchased from the nursery or seedlings you grew yourself.

The biggest question you'll want to answer is when you transplant. That depends on the plant. Some crops must be plated before it gets too warm (like lettuce). Alternatively, warm-season crops, like peppers, eggplants, and tomatoes, shouldn't be planted until the weather has warmed up, as they don't like cool temperatures. Soil temperature is also an important factor. Check the forecasts; that way, you'll know what to expect.

Preparation:

If the weather ahead looks good, it's time to start getting your garden ready:

- **Prep the Soil:** over winter, your soil might have compacted, so loosen it up. Use a fork to dig it over and aerate it. Get rid of debris and weeds, and dig organic matter into it about a spade's depth down. This will help it drain properly but retain moisture, allowing roots to dig down.

- **Warm Up the Soil:** do whatever you can to warm the soil up; place black weed suppressant or plastic across it and leave it there a week or two before you plant.
- **Don't Walk on the Soil:** place boards down to walk on, or make a pathway with something else. If you walk on the soil, it will become compacted, and roots will struggle to make headway. And when you water, it will just run off.
- **Starve the Plants:** that's not as bad as it sounds! One week before you transplant your seedlings, reduce the amount of water you give them and stop fertilizing. This will help them adjust to life in the great outdoors.
- **Harden Them Off:** you can't just take a seedling from a warm, sheltered environment and plant it in a cold, outdoor spot. You need to transition it and give it a chance to get used to the change. If you don't, the plant will go into shock and may die. A week or so before transplanting, place the seedlings outdoors in a shady, wind-free area, but not too shady, as they will need to feel the sun. Do this for a few hours daily, gradually moving them out of the shade and into the sun and wind. This will help them get used to their new, permanent environment.
- **Moisten the Soil:** during hardening off, the soil must be kept moist, as outdoor air can rapidly deplete the moisture in the soil.

How to Transplant:

Try to choose an overcast but warm day and plan to transplant in the early morning. This will allow your plants to settle into their new home without full exposure to the hot sun.

1. Test the soil to see if it is too dry or wet to dig holes; it should be moist, not drowning in water. Add a lot of water to the soil twenty-four hours before planting. This will keep the soil workable when digging holes, and the roots will be watered immediately when planted.
2. Level the surface before you begin to dig and plant.
3. Place the plants (either in containers or removed from them) on the soil before digging to get an idea of the layout.
4. Start with the plants farthest from the edges of the area. Dig a hole larger than the soil and roots in the container to place the

plant in.
5. Remove the plant from the container if you haven't already. Make sure you cover the soil side with your hand (do not damage the plant) and tap on the base of the pot - this will loosen the soil.
6. Pop the seedling into the prepared hole and fill in the hole with soil. Add a quarter-inch layer of soil over the top and tamp it down gently.
7. Settle the plant down by watering it generously. This will help acclimate the plant and fill any air holes left by digging and planting.
8. Give the plants time to fully acclimate (forty-eight hours) before you fertilize them. It is important to add fertilizer with phosphorous to help the roots take hold and for the plant to grow strong. Follow the directions on the label for adding the fertilizer.
9. If your planting bed is subject to hot weather, create ground cover to lock in the moisture by adding a layer of bark mulch on top of the soil.
10. Be mindful of the weather. If you have a bout of below-zero temperatures or a hail storm, protect your young plants by covering them to lock in the heat and keep out the elements. Remove the covering again when the weather returns to normal.

Never let the soil completely dry - you want to keep at least some moisture in the soil. Water it at ground level, which means don't hold the watering can or hose high above, as it can damage the plants - water daily until the plants are properly established.

Using Crop Covers

Plants that grow outdoors are vulnerable to the weather, the temperature, and many other problems that might come their way. This is why so many gardeners use crop covers to help protect their plants. Here are 10 reasons why you might want them in your garden:
1. Protect your young plants from digging garden pests like chipmunks, voles, and mice.
2. Speed up the germination process for direct-sowed seeds.
3. Shield tender plants from weather extremes and late frosts.

4. Shelter warm-weather crops from early frosts in the fall.
5. Keep the birds away.
6. Keep Mexican bean beetles, cabbage worms, hornworms, and other pests from laying eggs.
7. Cut down on the damage done by leaf-eaters like Colorado beetles, cucumber beetles, and squash bugs.
8. Stop deer from munching on your plants or brushing against your trees.
9. Keep cool-weather plants protected from the hot sun.
10. Keep your vegetables safe from rabbits, groundhogs, and squirrels.

Crop covers come in all kinds, each one offering its own type of protection. They are made from various materials of different sizes, some for single plants and others for several.

When Should You Use Crop Covers?

There is no specific time to use crop covers. You can use them as much or as little as you want, depending on what you're using them for. They can be used to protect from frost in early spring and late fall. If you want them to keep pests off your plants, use them throughout the growing season. And if you need to keep animals away, use them all year round. You get the picture.

The only other thing you need to know is not to wait until it's too late to use them. They are a preventative measure, not a solution for damage.

Here are the most common crop covers and their uses:

Row Cover

Row covers come in two types – plastic or fleece. Fleece is great for protecting tender plants from frost and is permeable, allowing moisture to seep through. Plastic creates a much warmer environment and is ideal for germination and bringing on seedlings.

You can extend your season with both row covers, but use plastic if your climate is colder. You will need to monitor them to ensure they don't overheat, and they will need to be secured down. Row covers allow you to get an early start on the growing season.

Best For: extending your season and protecting your crops.

Mulch

Mulch is incredibly versatile and is something every gardener should use.

It locks in moisture and heat when the sun is out but also insulates on cold days. A mulch layer will also limit the growth of weeds. And, if you use organic mulch, it can help condition the soil when it breaks down.

The important thing about mulch is not to use too much – no more than a 4-inch layer around a plant. If you use too much, the plants can suffocate. Mulch can be anything organic that can be shredded and broken down to provide a coverage layer (paper, bark, straw, etc.).

Best For: perennial plants that need a little protection from cooler weather.

Cloche

These make great temporary covers to protect tender plants from unexpected frosts. You might think your frost season is over or not due to start for another month, but Mother Nature often has other ideas! It's best to be prepared.

The only downside is that cloches are not cheap, so using them on large areas of plants might not be cost-effective.

Best For: protecting small numbers of young plants.

Cold Frame

Incredibly sturdy, cold frames are usually wooden-framed with glass panes and a hinged glass lid to allow easy access to your plants. Winter gardeners commonly use these and can help you keep some food growing all year round, even when it's snowing.

The key to success is ensuring your plants are fully grown just as the weather becomes icy cold. Plant growth is much slower during winter, so you want them to be as mature as possible before putting them in the cold frame.

Best For: fall/winter gardening in colder climates.

Chapter 11: Watering and Caring for Your Plants

Growing a successful garden isn't just about popping plants in holes and hoping for the best. Whether you grow from seeds or starts, your plants need a certain level of care, which includes watering. When your plants are mature, their care moves to another level.

Caring for your plants is the key to their survival.
https://unsplash.com/photos/EdscD_R28bM?utm_source=unsplash&utm_medium=referral&utm_content=creditShareLink

Seeds and Seedlings

Typically, you will need to water seeds and seedlings every day or two, regardless of whether they are indoors or in the garden. Make sure to water evenly so all areas of the soil receive moisture. You do not want water pooling or dry areas.

That said, this will all depend on the soil type, temperature, other heat sources, and so on. When the weather is hotter and drier, or if you are using a polytunnel or greenhouse, you may even need to water daily, if not more. Make sure you use a moisture meter to check the soil moisture regularly.

Here's a quick guide on when your seeds and seedlings should be watered:

- **The Top ½-Inch (1 cm) of Soil Is Dry.** Most seeds are sown just beneath the soil surface, and seedling roots are short, so they need the soil to be moist around them. If the top cm of soil is dry, they won't fare well. Do not let it get to this stage; it could halt germination and stunt growth. Soil is dry when its color is lighter. If you don't have a moisture meter, poke your finger into the soil without disturbing the seed or seedling; if it's dry, water it.

- **The Tray or Pots Feel Light:** lift them daily and feel their weight. The lighter they are, the drier they are. With a little time, it won't take you long to learn when your plants need watering just by picking them up.

- **Check the Plants:** small seedlings are sensitive to changes in water; if not enough, they will start drooping. If you see seedlings like this, water them, but don't overdo it.

Direct-sown seeds and seedlings are a little easier to care for, and you have a bit more leeway. Pot-grown plants tend to dry out quicker than plants in the outdoor soil, and moisture is in shorter supply, whereas plants grown in the outdoor soil have access to a much deeper supply of water, which also prompts a plant to grow a deeper root system. They also benefit from dew in the mornings and rain showers.

As seedlings grow and age, they aren't quite so greedy for water. A week or 10 days after germination, you can get away with reducing the watering to every other day, and as they continue to grow, you can reduce it even further, so long as when you do give them water, it's a deep watering.

Under and Overwatering Your Seedlings

Obviously, if you don't provide enough water, your seedlings will dry out and can die, especially in hot climates. Older plants can be revived, even if they are very dry and a little wilted, but younger plants don't have the resilience to survive without water, even for a few days.

Another problem with under-watering occurs if you use peat moss in your growing soil. When peat moss dries out, it won't soak in the water; instead, it just runs off.

If your plants have dried out, water them as soon as possible; you might be lucky enough to catch them in time. If you have used peat moss and it is dry, soak it in a tray of water until it is rehydrated.

What if you overwater your seeds and seedlings? That's okay, isn't it? Well, no, it isn't. Many people make the mistake of flooding their plants with water when they've dried out, but this can lead to its own set of problems, including:

- **Root Rot:** when the soil is saturated with water, the roots can rot out.
- **Drowning:** yes, you can drown your plants because they can breathe. Water can fill the holes in the soil, stopping air from getting in, and the plants will drown.
- **Mold:** mold loves the damp, and it's fatal for young plants and seeds.
- **Damping Off:** this is a fungal disease that affects overwatered seedlings.
- **Insects:** some pests love the damp, and they'll attack young seedlings, which don't have the strength to survive.

What to Do If You Overwater Your Seedlings

If your seedlings are in trays, move them somewhere dry, airy, and sunny to help them dry out. If they are in the garden, you can only wait until the soil dries out and pray you don't get heavy rain.

The Proper Way to Water

Watering your plants isn't an exact science, but it must be somewhere near the mark for them to thrive. There are two ways to do your watering:

1. Bottom Watering

This uses the principle of capillary action. Water is taken from very wet to drier areas.

Place your pots in a shallow tray and fill the tray with water. Leave it be for a couple of hours; by then, the soil will have soaked up what it needs - you can check this with a moisture meter. If it is still dry, leave it for a bit longer. When the soil is moist enough, get rid of the remaining water. Keep an eye on things; your plants may be incredibly thirsty and go through the water quickly. If the tray dries out quickly, add more water.

This is one of the best ways to water seedlings as it is gentle, and the soil will only take in what it needs, which means no chance of overwatering.

2. Top Watering

Self-explanatory. This means you water from above. However, how you water from above depends on whether your seedlings are growing indoors in containers or outdoors in the ground. Those indoors are in lighter soil, which could be washed away if you are not careful, or you could break the seedling stems.

Here are the best top-watering methods for pot-grown seedlings indoors:

- **Mist:** use a spray bottle to mist water onto the seedlings, usually once a day or more. This will only water the soil surface and won't soak through the soil. You only need to do this until the seeds have germinated and started showing signs of growth; then, they'll need more water.

- **Lightly Sprinkle:** a watering can with a good hose on it offering a fine, light spray will work for this. Alternatively, make holes in the lid of a water or soda bottle and fill the bottle with water. If you use a watering can, use an indoor one -they're usually smaller, with thin spouts, and much gentler.

Watering your outdoor seedlings is as simple as using a hosepipe or watering can. However, the same principles apply; if they are young, don't squirt them with force. A gentle spray will do the trick. Alternatively, you could use a drop irrigation or soaker hose setup. Once the system is set up, simply attach your hose and leave it - the water will go deep into the ground around the roots.

Why Is Watering the Right Way So Important?

As mentioned earlier, it isn't an exact science, but it's also not about just tipping a load of water onto your plants. Understanding how plants use water takes time to learn, and other factors are also at play, including temperature, weather, soil texture, time of year, and time of day. You need to pay attention to all these factors because your plants will need different amounts of water at different times.

However, while you are getting to grips with your garden, there are some tips you can follow to help you:

1. **Water at the Roots:** water should be at soil level and applied until it has soaked the whole root ball. Don't forget, the root system on a plant can go a long way, so assume it is as wide as the actual plant and at least a foot down, if not more. That's where soaker hoses are best, as the water goes down through the earth to root level; 20 minutes and your plants will have all the water they need.

2. **Check the Soil:** don't water because you think you should; your plants might not need it. Use your fingertip to probe a couple of inches down into the soil; if it's dry, it needs water. Alternatively, use your moisture meter.

3. **Water Early:** the morning is the best time for watering, as the leaves can dry out during the day. If your plant leaves are constantly wet, there's a good chance that fungal disease will take hold. If early morning is not a good time for you, leave your watering until late evening, when the sun is not as strong.

4. **Water Slowly:** if you blast water at dry soil, it will run off and not sink in. Start slowly and build up; when the soil is moist on top, it will be better absorbed deeper down.

5. **Make It Count:** again, soaker hoses and drip irrigation systems are best. The water goes only where needed, and there's no waste. Also, early morning or late evening watering minimizes water loss through evaporation. For certain crops, the best companions are those that shade the ground, keeping moisture in.

6. **Not Too Much:** water isn't the only important thing to a plant; it also needs oxygen. Allow the soil to dry out a little between watering, especially with container plants. A deep watering once or twice a week is far better than daily watering.

7. **Don't Let Them Dry Out:** when the sun is at its highest and hottest, plants often wilt a little; this allows them to conserve some moisture, but you should see them perk up when the day cools down. If you don't see this, your plants have gotten too dry. This damages some of the tiny projections on their roots, and the energy it takes to regrow these should be going into plant growth. That could result in stunted growth and poor harvest.
8. **Use Mulch:** using organic mulch around your plants can help retain moisture longer as it stops water from running off or evaporating. However, don't use too much mulch, as it can stop the water from getting to the roots.

Fertilizers

Plants need healthy soil to grow properly and produce fruit and flowers. All plants take nutrients from the soil. Some need more nutrients than are available. It's sort of a chain reaction:

- You feed the soil with nutrients
- Those nutrients feed the plant
- Plants feed us

Plants need three primary nutrients: nitrogen, phosphorus, and potassium. They cannot absorb nitrogen from the air, so they must get it from the soil, and if there isn't enough, fertilization is needed to boost it. Potassium exists in the ground but is usually far deeper than the roots go and isn't available to them. Phosphorus is also available, but only in certain rocks. The only way a plant can access this is if it is water soluble. That's why we need to give our plants a helping hand.

Organic fertilizers, such as compost and animal manure, are best, but you can also purchase fully organic fertilizers. Later, you'll learn how to make your own, but for now, refer to Chapter 9 for details on how to add fertilizers to your soil and plants.

Pruning and Deadheading

Experienced gardeners and newbies must follow routine maintenance to raise healthy plants, including pruning and deadheading. Not every plant will require this kind of care, but you must understand what to do when necessary. You might think pruning and deadheading are the same, but they are different methods and tend to be done at different times during the season.

Pruning

Pruning should be done regularly on some plants and is the practice of removing branches and foliage. You can use pruning to remove dead or diseased parts of the plant, trim shrubs into new shapes, or encourage growth.

Pruning promotes fresh growth, new flower buds, and a healthy plant. If you have old shrubs in your garden, pruning can give them a new lease of life. Look at it this way: just like you need to get your hair cut now and then to freshen and tidy it up, your plants need the same thing.

How to Prune

This is actually quite simple. Cut off what isn't needed using a sharp pair of secateurs or pruning shears. If you thin a plant's foliage, cut off up to a third of the stems. If you are pruning to cut a plant's growth back, don't cut off the stems, just the offending growth.

Annuals and perennials should be pruned once the first flowers have shown themselves and stop when the growing season ends. However, you must research because each plant is different and requires different pruning techniques and timings. Some plants have seasonal requirements and can only be pruned at certain times of the year.

Deadheading

Deadheading is an intuitive gardening process. Dead flower heads can still remove nutrients from the soil but won't grow. Removing them gives more food to the other flowers and flower heads. All you need to do is remove the dead heads.

You will also have a nicer garden when it is not decorated with dead flowers. When you remove dead flower heads, you might notice your garden bloom more and become more colorful now that the nutrients in the soil are not being wasted.

How to Deadhead

Simple. Just snip off any dead or faded blooms. Cut them off just above the first set of leaves below them. You can do this as often as you want or stick to doing it once a season; however, the less you do it, the fewer blooms you will have.

Not all plants need to be deadheaded. Typically, those that produce plenty of flowers, such as marigolds, roses, petunias, salvia, etc., benefit from deadheading. However, if they only have a single bloom, they won't thank you for cutting it off.

Chapter 12: Troubleshooting Common Companion Planting Issues

No garden is without problems and challenges, but understanding the problems and how to resolve them will help you keep your garden in good condition. This is especially true where companion planting is concerned, but thankfully, as it's a centuries-old technique, there's plenty of advice on potential problems. Here are the most common ones:

Insufficient Spacing

Many gardeners make the mistake of planting things too close together. You may not realize it then, but you'll soon see it when your plants mature to full size. You want your companion plants to do their job without crowding your crops.

How to Avoid:

Be sure to plan ahead. You do not want to allocate space based on the size of seedlings or seeds. Consider how big the plant will become, and space the plants out early - even if it looks like too much space when planting. If you are unsure how much space you need, always err on the side of caution and give more space than needed.

Competing for Water

This will depend on your soil type and its ability to hold water. If water is in short supply, hardier, deep-rooted plants will take the water and leave nothing for the others, a problem that also happens when plants are fighting for space.

Shallow-rooted crops struggle with water because the top of the soil dries out first, and they don't have taproots to help them get water. When companion planting, ensure deep-rooted plants can't take water from shallow-rooted ones.

How to Avoid:

Keep your soil moist. Use soaker hoses and mulch to keep the water consistent throughout.

Competing for Nutrients

Some plants need many nutrients, such as brassicas, tomatoes, cucumbers, squash, and peppers. The right nutrient levels allow them to produce a long season of fruit. However, they can steal all the nutrients in the soil, leaving nothing for other plants. If there are insufficient nutrients, the crops will suffer.

How to Avoid:

Pair your crops per their nutritional needs, i.e., don't plant companions that need the same nutrients as your main crops. Make sure to add plenty of organic fertilizer throughout the season to keep the levels up, particularly slow-release fertilizers.

Shading Your Plants

Companion planting offers fantastic benefits, but if one or more start overgrowing, they can shade the others out, and there go your benefits. Sunlight is the fuel required for plant growth and photosynthesis; if plants have to compete for light, it won't end well. Let's say you allow your cucumbers to grow along the ground rather than upright (not a good idea); taller plants will cut out their light and stop them from growing and fruiting properly. Likewise, tall tomatoes can crowd the light out from bush beans. However, some shade also benefits some plants, such as spinach and lettuce.

How to Avoid:

Almost all plants and flowers need sunlight to grow. When companion planting, ensure your taller plants are not providing too much shade. It can be tempting to over-plant a companion without realizing the damage it will do. For vine-like plants, provide support to raise them to get the sunlight they need. Study your garden before planting to see where the sun hits each day; this guides you better in terms of how to shade your plants.

Allelopathic Companions

Plants are a bit like people; they don't all get on together. Some plants stop others from growing, and this leads to certain disasters. Allelopathic plants produce chemicals from their roots that suppress growth in neighboring plants – basically, only the strongest will survive, and these plants are only interested in themselves.

How to Avoid:

Be careful with your companion planting and ensure no allelopathic plants exist. Here are some of the worst combinations:

- **Alliums or Mint with Asparagus:** Alliums and mint are both good at pest control because of their volatile oil production but can reduce the growth of companion plants
- **Onions and Beans:** They do not work well together and will limit growth, especially from seed
- **Sunflowers and Potatoes:** Sunflowers limit the growth of potatoes with the chemicals they release; sunflowers also produce too much shade.
- **Fennel and Almost Everything:** The compounds released into the soil by fennel stunt the growth of almost everything around them

Different Soil Requirements

Changing the soil composition from one spot to another is almost impossible, and not all plants thrive in the same conditions. Some plants prefer alkaline soil, while others prefer acidic soil. Some plants can thrive even when the soil is changed, starting as a great companion before becoming an enemy. Plant these near one another, and one of your crops won't grow properly.

How to Avoid:

Understand your plants' soil and pH requirements, and only plant those companions that work with your crops, not against them.

Bad Timing

Timing is important with companion planting. The right companions will complement one another perfectly throughout the entire growing season. For example, if you plant tomatoes, fill the rest of the bed with lettuce or radishes. They are quick plants that will be fully harvested when the tomatoes are fully grown.

Some plants must be established before their benefits as a companion become apparent. For example, if you plant corn and cucumbers together and want the cucumber to use the corn as a trellis, the corn must be at least a foot or two tall before you plant the cucumbers.

Lastly, flowering must be considered. The flowers offer the best companion planting benefits, so while the foliage on many plants does a good job at repelling pests, their flowers bring beneficial insects and predators in.

How to Avoid:

Look at the DTM (Days To Maturity) of every plant. Maturity is usually the time taken to reach the first harvest – some plants will continue fruiting all season. Synchronize your planting times so everything reaps the benefits, and also experiment; try staggering plantings to see what works.

Aggressive Companions

Some plants are overly aggressive and really shouldn't be planted beside vegetables:

- Bamboo
- Bee Balm
- Blackberries
- Clover
- Jerusalem Artichokes
- Mint

- Morning Glory
- Rosemary
- Thyme

These plants can be stunning but grow so fast, choking everything else in their path. Some will spread fast underground – bamboo, rhubarb, mint, etc., and pop up all over the garden, bypassing any barriers you might have put up.

How to Avoid:

Don't plant these anywhere near your vegetable beds. Plants like rosemary, mint, rhubarb, and thyme can be planted in containers. That way, you get their companion plant benefits without the downsides they come with.

Haphazard Plantings

You might think a haphazard, messy planting shape is fun, but rows are there for a reason. They are visually appealing and make it easier to tend to the plants. Watering (as does weeding) becomes hassle-free, and you can easily see which plants are growing and which are not.

How to Avoid:

Keep the plants in rows. This is easy to do with some forward planning and will make life much easier in the growing stage.

Different Maintenance Requirements

You should already have discovered that the real secret to successful companion planting is to plant like with like. That means planting those with similar requirements together. If you plant companions with different requirements, things will almost certainly go wrong. Some plants require you to add additional soil during the growing season, and if you plant them with low-growing plants, the low-growing plants will be covered.

How to Avoid:

Be sure you know what the plant needs before planting it. A little research can go a long way toward successful plants and reduced problems.

Wasted Space

If you only have a small garden space, companion planting is a fantastic way to make the most of your space and increase your yields. However, not understanding how to time your crops, when and what to trellis, and how to space everything could lead to a lot of wasted space.

How to Avoid:

Map all the bare spaces in your garden and determine how to fill them. For example, when you plant young tomatoes 1 to 2 feet apart, you have a lot of empty space. That could be filled with fast-growers, like lettuces, spinach, or radishes. By the time the tomatoes reach maturity, those crops can be harvested. If you put another slow-grower, such as peppers, into a bed, interplant with basil or scallions to fill in the empty gaps.

Attracting the Same/Similar Pests

When two plants attract the same kind of pests, planting them together is sure to end in disaster. With a choice of crops to attack in one place, pests are more likely to settle in, breed, and destroy your garden. This happens because the crops look and/or smell the same. For example, two brassica family members, cabbage and kale, attract aphids in large numbers. Plant them together, and you have a huge problem on your hands. By interplanting with onions, sweet alyssum, or calendula, you can stop those pests from jumping between your plants; even better, plant some nasturtiums a distance away. These are trap crops that attract the aphids away from the crops.

How to Avoid:

Select your companion plants carefully in terms of the pests they attract. Do not plant the same family of plants too close to each other, such as:

- **Brassica Family:** broccoli, Brussels sprouts, radish, cauliflower, kale, mustard
- **Cucurbitaceae Family:** squash, cucumbers, melons, zucchini
- **Solanaceae Family:** potatoes, eggplant, peppers, tomatoes
- **Amaranthaceae or Chenopodiaceae Family:** chard, beets, spinach, quinoa

If you have to plant them close together, plant companion plants that are excellent at deterring pests, or your garden will be overrun.

Too Many Companions

Companion planting is fun, but going overboard when you are a newbie is easy. When you add too many plants to one bed, you create more problems than you solve. The garden becomes overgrown, and you will struggle to tend to any of your crops; not only that, but you also won't be able to see which companions are working.

How to Avoid:

Having a diverse garden is wonderful, but aim for simplicity when you first start. Try a couple of combinations per garden bed to learn what works together and what doesn't. As you gain experience, you can experiment a little more.

Not Using Groundcover

Companion planting isn't just about keeping pests away; certain plants act as ground cover to keep moisture in and weeds down and as a habitat for beneficial insects. These plants may not always look colorful or smell amazing, but they do a fantastic job in the garden.

Ground-cover plants can:

- Limit weed growth
- Provide ground cover to keep heat from the soil
- Elevate other plants to reduce rotting when fruits are in contact with the soil
- Replace a layer of mulch
- Lock in moisture in the soil for long periods
- Enhance the soil with root growth

How to Avoid:

If you do not have ground cover, planting a low-growing plant can add that to your growing area. Micro-clover is great and will also add nutrients to the soil; creeping thyme is another hardy companion. Pair them with taller plants that provide shade where necessary, and half your work is done by the plants.

Companion planting is not complicated, but you do need to do some planning to keep the mistakes to a minimum. Before you decide on your companion plants, ask yourself these questions:

- Have I given each plant enough space to reach full size?
- Will these plants fight for nutrients and water? Are the fertilization needs similar?
- Will one plant grow bigger than the other and shade it? If yes, does the smaller plant benefit from a little shade? Does it need a lot or a little?
- What does the plant do for pests? Will the same pests attack both plants?
- Is one of the plants allelopathic? Will it attack the other one?

If you don't create a solid plan, you might end up with a disaster zone in your garden. However, if you do try a combination that doesn't work, don't worry; we've all done it! If it causes too many problems, pull up the companion plant and replant it elsewhere in the garden.

Companion planting takes time, but keeping a notebook can help; note how your plants work together or cause problems so you know what to do next time.

We've finally got there - it's harvest time!

Chapter 13: Harvesting Your Organic Companion Planting Garden

You've put in all the hard work, and now it's time to reap the rewards - that succulent harvest just waiting to be picked. But how do you know when it's the right time?

That is one of the most common questions gardeners ask, and that's because most inexperienced gardeners have a preconceived notion of how their fruit and veggies will look - just like it does in the grocery store, right?

We also pick too soon, impatient to get our hands on what we've grown, or we go the opposite way and leave it too late.

First, here are some tips to help you with your harvest:

1. Get in Your Garden Daily

When your harvest starts ripening, it can all happen simultaneously; that's why you should be out there every day. If you don't wander your plants regularly, you will miss early ripening produce and leave it to rot. This attracts pests and diseases, which can soon wipe out your garden.

You don't want any of this, so check your plants every day. When you spot ripe fruits and veggies, pick them. This does two things - it gives you delicious food to eat, and in some cases, it can encourage a plant to continue fruiting - tomatoes and cucumbers are two examples.

2. Pick Small

How often have you looked at your zucchini and thought, *I'll just let it get a bit bigger?* Before you know it, it's huge, which is not good. If you let your veggies grow too big, they lose flavor. Smaller vegetables taste better, are more tender, and don't have too many seeds.

That said, if you do find a huge zucchini or a tomato that looks like it ate an entire box of growth hormones, don't chuck them; you can still use them in your cooking.

Pick small and enjoy more flavor while encouraging your plants to continue producing.

3. Gently Does It

You can involve the kids in harvesting, but they must be gentle – it doesn't take much to bruise fruits and vegetables. They must be picked off the plant gently and placed in the harvest container even more gently!

It's not because bruised produce doesn't look nice; it can hasten rotting, leading to a much shorter lifespan of your harvest. If you do bruise any, you need to use them immediately.

4. Make Sure Your Harvest Containers Are Large Enough

Ensuring you place your harvest into large enough baskets will lessen the chance of bruising, which means you can bring in a bit more each time. 5-gallon buckets are ideal for beans, and large baskets (bushel-type) are best for larger plants, like squash, eggplant, cucumbers, etc.

You could use a clothes basket if you have nothing else or even a wheelbarrow, but be careful with softer fruits.

5. Look Where You Are Walking

Make sure you have clear pathways between your plants.
https://unsplash.com/photos/1_yycyoMT6g?utm_source=unsplash&utm_medium=referral&utm_content=creditShareLink

This is important, especially if your garden is well-grown. Create clear pathways between your paths, or be careful where you walk. You may step on smaller plants or lower fruits and vegetables. Not only will this damage the crops, but it can also open up the doorway to pests and diseases, quickly wiping out your harvest. If you do accidentally step on a vegetable or fruit, pick it up and dispose of it immediately.

6. Keep Track

Keeping up with everything is difficult when you have a busy garden full of different plants. You need to know what crops you planted, each variety, the time to harvest, and what you should look for in a harvest.

Keep a garden journal – you can buy specially designed ones or use a notebook – and write down everything. When you know all this information, you know when you need to start checking for harvest. You can track the plant growth in your planner and note down the start of the harvest season so you are ready to harvest – but also know not to be away from your garden during this period. It can be tempting to harvest too early, and it is not uncommon to harvest too late, but by noting down when you should harvest based on the plat type, you can avoid these common problems.

7. Monitor for Disease

When the seeds are sprouting and right before harvest – these are two critical times to check for any possible disease. Check for misshapen leaves, discoloration, or dead parts. Check your plants regularly; it can help you pick up on early signs of trouble and take steps to correct it. If you do find disease or pests, look at your companion plants and decide if you could plant something different next time to stop the same thing from happening again.

8. Be Realistic

Don't base your expectations of your plants on what you see in stores or on seed packets. For example, the heads on home-grown broccoli don't tend to be as large as those from the stores. If you have unrealistic expectations, you won't know when to harvest; you'll be looking for something that doesn't exist. Leave them too long, and they will soon start to rot.

Knowing what to expect from your harvest is the easiest way to know when to pick it.

9. Quickly Harvest Stems

The stems that are not producing fruit or flowers should be removed as quickly as possible to allow more nutrients to go where needed. Leafy vegetables and herbs should be harvested early to lock in the flavor - leave them too long, and the flavor will deteriorate as they use more nutrients.

10. Allow the Fruits to Hang

Those plants that produce fruit, such as apples, tomatoes, peppers, etc., shouldn't be picked too early, or they won't be fully ripe. Leave them to ripen fully before you harvest them; this comes back to knowing your varieties and what to look for.

Let's look at some of the most popular crops so you get an idea of the best time to harvest them.

Popular Plants – Harvest Times and Methods

Most plants can be harvested without the need for special tools, just a pair of gloves and a basket. However, in some cases, you can use pruners or a small knife to help you out. Here are some of the most popular plants you are likely to grow and tips for harvesting.

Herbs

Learn what your herbs should look like when they are ready for harvest. Then, you should cut them back as often as possible and store them in the refrigerator on dry paper towels to soak up the moisture. Alternatively, hang them somewhere cool and dry where they can dry out before you store them for harvest. You can also chop herbs and place them in ice molds with oil. They can then be used later individually.

Tomatoes

Tomatoes come in various colors and sizes depending on the type. A tip here is to look at the seed packet if you grew your own, or the nursery label may have a picture of a mature plant on it. The fruit should be firm but gives a little when gently squeezed.

Ripe tomatoes should come off the stem easily. Pull on them gently; if they come off, they are ready.

Peppers

Peppers will also change color as they ripen, but many varieties can be picked at any color. Expect to see green, yellow, orange, and red. Peppers will sweeten over time, so the longer they grow, the sweeter they taste. Be sure to check the harvest time of the peppers you are growing. You can grow them for too long if you are not careful.

When harvesting, cut the peppers from the plant rather than pulling them off. Hold the stem and twist the pepper if you don't have a knife handy.

Lettuce

Lettuces can generally be harvested when the leaves are about 4 inches long, depending on the variety. Try to pick them while it is still cool outdoors; in extreme heat, lettuce can start to produce seeds, giving the leaves a bitter taste.

Leaf lettuces should be harvested from the outside in, while head lettuces (iceberg, for example) should be sliced off at the stem. Most lettuces are known as cut-and-come-again, which means they will continue to produce.

Green Beans

Beans are another crop that continues to produce when picked. You can enjoy a good harvest all season with just a few bean plants. When you see blossoms on the plants, start checking; pick the pods young for a sweeter bean. Don't leave green beans for too long, or they will no longer be tender and soft – check the seed packets for harvest times.

Do not pull on them too hard; you might end up pulling the whole plant out. Use scissors or secateurs to snip them off. Do not harvest beans in the morning as they will likely still be wet and dewy, which can spread disease.

Green beans are the perfect example of natural fertilizer. Once the plants are finished providing beans, chop them down and leave them on the ground. Let them start to die off and dig them into the ground, roots and all; this is the perfect way to boost your soil with nitrogen.

Peas

This is a trial-and-error crop in terms of harvesting. Check the peas regularly by opening a pod and tasting the peas. If the peas are the size you want, continue harvesting; if not, leave them a bit longer. Again, once the harvest is done, dig the pea plants back into the ground for a boost of nitrogen.

Melons

Seriously, testing a melon for ripeness is as simple as thumping it. If it sounds hollow, it's ready. If you don't want to do that, have a good sniff; most melons give off a sweet aroma when they are ripe.

Harvesting is as simple as snipping the fruit off the vine.

Watermelon

Check the part touching the ground to see if your watermelons are ripe. The melons should be green and stripy and have a yellowish spot on them where they lay on the ground. The fruit is not ripe if that spot is white or light brown. Again, simply snip them off the vine.

Cucumbers

Knowing the variety you planted can help you determine the size of the cucumbers. When they get to that size, harvest them. Leave them too long, and they will produce a lot of seeds inside and be bitter. Do check your plants thoroughly; they are quite leafy, and you might miss one or two that grow into massive fruits.

Harvesting is done by tugging and twisting gently to remove them. However, you can also use secateurs or scissors to cut them off.

Corn

When your corn starts to form ears, give them a gentle squeeze. A husk covers the corn cob, but you can feel the corn beneath. When the strands of husk begin to dry, check a corn kernel if you have access to it. Squeeze the kernel between thumb and forefinger - if white sap emerges, your corn is ready.

Corn husks are easy to remove from the stem when ready.

Root Vegetables

Keep track of all your root vegetables so you know when to harvest, as the harvest time is tricker to determine by examining the plant alone. When ready to harvest, gently pull one plant to check the size of the vegetables. With carrots and beets, you can thin them out earlier in the season. This means pulling every plant out - you should have fully formed baby carrots or beets, which are delicious. This leaves the rest of the harvest room to grow - two crops for the price of one.

Garlic

Check the tops of the garlic to determine whether it is time to harvest. The tops will turn from green to brown when the garlic is ready. Don't do any prep or cleaning other than hanging the garlic to dry once it is removed from the ground.

Eggplant

Eggplants can be quite bitter if you let them get too big. Instead, harvest them when small and purple with a nice sheen; they should also be firm to the touch. Do not pull eggplants off the plant; you will destroy the entire plant. Instead, snip them off and allow your plant to put its energy into producing more fruit.

Onions

Onions take a long time to mature; you usually won't gather in your harvest until the end of the growing season. Like the garlic, look at the tops; you can harvest your onions when they have died off. However, if you see an onion with a long, thick stem in the middle and a flower head, cut it out and harvest it immediately - left in the ground, it will go tough and may rot.

Harvesting is simple: pull them from the ground. If you are storing them over the winter, they must be cured. Lay them in a single layer with space around each one and let them dry. If your weather is warm and breezy, they can be laid on top of the earth to dry.

Potatoes

Potatoes are ready for harvest when the leaves have turned yellow and died back. To see what's grown, root around in the earth at the base of the plant; alternatively, just let the plants die back and dig up the spuds.

Harvesting depends on your growth method. If you grow them in mounds or trenches, carefully remove the plant and dig around in the soil for the potatoes. If you use a fork, be careful not to stab any. If you damage any potatoes, these must be used quickly and cannot be stored, or they will rot - one rotten potato can cause all the others in storage to rot. If you grow your plants in potato bags, tip them onto a tarp and sift through the soil.

Carrots

Carrots will be ready to harvest in around 7-8 weeks, depending on your climate - smaller varieties of carrots will not take as long. When you are ready to harvest, simply pull them from the ground. Carrots are robust, and you can leave them in the ground until you are ready to eat them instead of pulling and storing them – just be careful to watch for frost.

However, unless you are growing them in a polytunnel or your climate is quite warm, you cannot leave them in the ground over the winter; they will freeze into the soil.

Kale

Kale is a simple plant to harvest.
https://unsplash.com/photos/M8bpp4qQZGg

Kale is a simple plant to harvest; just wait for the leaves to be large enough. This will normally be around 10 inches long, although you can pick them slightly smaller if you prefer.

Harvest the outside leaves first; kale plants will grow new leaves and produce large quantities throughout the season. These are usually hardy and can survive winters on the ground.

Summer Squash

Summer squash matures quite quickly, so long as they are pollinated successfully. The fruits are quick to grow and can be harvested at the size you prefer; the bigger they are, the more seeds they will have. Most summer squash varieties take about two months to mature.

You don't want to pull squash, or you can damage the flower; use a blade to cut them from the stem (always use a clean knife to limit the spread of plant diseases). If you do damage the stems, they might not produce any more squash.

Snap Peas

This variety of peas can be harvested in around seven weeks. Harvest them early before they become tougher and more stringy. Check these daily around harvest time.

Do not tug the pods off, as you will damage the plant. They should snap off easily.

Brussels Sprouts

Sprouts are usually ready to harvest when they are around one to two inches in diameter. These are a slow-growing crop, though, so be patient. You usually won't harvest until the end of the season.

Pull sprouts off as needed or harvest the entire plant to keep the pests off. They can be blanched, frozen, or stored in the refrigerator for up to two weeks.

Beets

Beets may be harvested as babies or mature. The average maturity time depends on the variety you grew, so check the seed packet. If you leave them too long, they will become tough and won't taste very nice. Like carrots, they cannot be allowed to freeze in the soil, but a touch of frost won't hurt them.

Harvesting is no more difficult than pulling them from the ground. You can also eat the leaves; add them to salads or stir-fry them with a little salt and garlic.

Spinach

Most spinach varieties will bolt very quickly, so keep an eye on them and harvest the leaves as often as possible – baby spinach leaves are lovely.

Pull the plant and use the leaves if it looks like it will bolt. Otherwise, just pick off leaves as you want them, and they will regrow.

Keep a constant eye on your garden; you will soon learn what's ready to harvest and what isn't. It also means you won't lose plants to bolting or end up with bitter fruits because you left them too long.

Bonus: Organic Fertilizer Recipes

Organic fertilizers are available in almost every nursery or garden center, but why buy them when you can easily make your own? Our final chapter provides you with some easy organic fertilizers to make using what you would normally just throw away.

Grass Clipping Tea

Fresh grass clippings are full of nitrogen, and you shouldn't add too much to your compost pile. However, you can use them as mulch - don't put them too close to your plants, as the grass is acidic and can burn them. Layer the clippings no more than two inches deep; any more, and they will flatten into a wet mess that won't allow oxygen through and can cause your plants to go moldy. You can also make a nitrogen tea to feed your plants:

1. Take a five-gallon bucket, and one-third fill it with fresh clippings. Fill the rest of the bucket with clean water.
2. Leave it for two weeks, stirring now and then.
3. Strain the grass from the liquid and mix it one part grass tea to five parts water - it should be like a weak tea. This should be applied at the soil level, not on the leaves.

Manure Tea

If you can get hold of fresh livestock manure, you can make a tea your plants will love:

1. Fill one-third of a container with manure and two-thirds water.
2. Leave it for three days, stirring now and then
3. Strain the tea and throw the manure on the compost heap
4. Dilute the liquid with water until it turns a pale brown-yellow, see-through liquid.
5. Again, this should be applied at the soil level, not on the leaves, especially on spinach, lettuce, and brassicas.

Dandelion Tea

Dandelions are full of potassium that plants need for photosynthesis, and you can use the entire plant to make tea:

1. Harvest your dandelions – the tops of the whole plant; it's up to you. Do NOT use any that have been sprayed with herbicide.
2. Put a good bunch of dandelions into a five-gallon bucket and top it up with water.
3. Put a lid on and leave it for three to four weeks, stirring now and then. As the dandelions break down, there may be a smell, and the water will go black.
4. Strain it and discard the dandelions onto your compost heap.
5. Dilute the tea to a light color and apply at the soil level. This will encourage the plant to flower and produce fruit.

A word of advice on dandelions: avoid spraying them with chemicals, and don't pick them or mow them down too early in the season. They are often the first food source for pollinators like bees; if you kill them off, the bees can't feed.

Banana Peel Tea

Another great way to feed your plants potassium is to make tea from banana peels.

1. Gather enough banana peels to fill a container. If you don't eat that many bananas, chop the peels of what you do eat and freeze them. When you have enough, you can dump them in the container.
2. Fill up the container with water and leave them for one or two weeks

3. Give it a good stir, then strain it; the remaining peels can go on your compost heap, and the tea can be diluted one part to five parts water until it is a light color.
4. Feed at the soil level until needed, or use it as a foliar spray to deter aphids.

Crushed Eggshells

Eggshells are full of calcium and can help raise soil pH. However, if you just throw them in your garden or the compost heap, they won't break down for years, meaning calcium isn't readily available. The quickest way to solve this is to crush or grind them:

1. Save your eggshells and spread them on a baking pan. When the oven has been on, place the pan in and dry the eggshells for a few minutes until brittle. This will also kill off salmonella bacteria.
2. Grind the eggshells to a powder consistency or crush them down into a fine consistency.
3. Apply as a top dressing around plants that need calcium, dig them into the soil, or add them to one of the fertilizer teas.

Coffee Grounds

Coffee grounds usually get tossed in the trash but make a great fertilizer packed with magnesium, nitrogen, and potassium.

1. Spread your coffee grounds on a tray and let them dry out.
2. Once dried, you can sprinkle them sparingly around your plants.
3. You can also add these to one of the teas mentioned above.

These are perfect for plants that love acid, i.e., rhododendrons, azaleas, blueberries, and roses.

Epsom Salts

Most people have a box of these lying around; if not, they are readily available. They are full of two secondary nutrients – sulfur and magnesium.

1. Add one tablespoon of Epsom salt to one gallon of water and stir it thoroughly to dissolve the salt.

2. Use this to water your plants once a month throughout the season, especially tomatoes, potatoes, peppers, and roses.

Vinegar Fertilizer

Vinegar adds acidity to the soil. If you have plants that need acidity-rich soil, white vinegar is one of the best things to add to your fertilizer. Vinegar is great for house plants as it does not harm any children or pets.

1. Add one cup of white vinegar to a gallon of water and stir it in
2. Water your plants with this once every three months

Never use undiluted vinegar on your plants, as it will kill them.

Compost Pile

Making a compost heap is one of the best ways to feed and fertilize your soil. You can do this directly on the ground, build or buy a proper compost bin, or just use a bin. Throw in all your vegetable and fruit scraps, some grass clippings, and any other compostable material – make sure you add cardboard, newspaper, or shredded paper, as this balances the compost and helps it turn quicker. Add a little water now and then and turn it to speed up the composting. You can also purchase rotating composters – add the material, shut it, and turn the handle.

Although it takes a while to make compost, when it's ready, it will feed your soil with microorganisms and nutrients that help feed your plants during the next season.

Mix and Match

Commercial fertilizers are usually a combination of nutrients, and you can emulate this in your own home:

- When you make your grass-clipping tea, add a tablespoon of Epsom salts and some banana peel to the container
- Combine your dandelion and grass clipping teas, and add a healthy dose of crushed eggshells

Get creative; have fun and make lots of notes so you know what works for next year

Conclusion

Whether you knew nothing about companion planting and gardening or were already experienced, I hope you now have more knowledge to put to good use.

Companion planting is an important part of gardening. It's not about making your garden look pretty; although done correctly, it can have a stunning visual effect. It's about gardening organically, about using plants to keep pests at bay and control weeds without using chemicals. It's about helping the soil structure, feeding it with nutrients, and helping other plants to thrive and produce a bountiful, healthy harvest.

This easy-to-read guide has provided you with information on everything you need to know, including:

- What companion planting is and how it all started
- How to plan your garden
- What plants make good companions, and what don't
- How plants help each other
- The difference between starting from seed or buying starts
- How to plant your garden
- How to care for your garden
- How to harvest your bounty
- How to make and use organic fertilizers

If you are a beginner, this book has hopefully awakened your passion to get out in the fresh air, get your hands dirty, and produce a gorgeous, healthy, organic garden.

Here's another book by Dion Rosser that you might like

References

"11 DIY Homemade Plant Fertilizers (with Recipes)." *Gardening*, 21 May 2022, https://gardening.org/homemade-plant-fertilizer-recipes/ .

Andrychowicz, Amy. "Seed Starting 101: The Ultimate Guide to Growing Plants from Seed." *Get Busy Gardening*, 16 Mar. 2017, https://getbusygardening.com/growing-seeds/

Angelo. "What Is Companion Planting and How Does It Work?" *Deep Green Permaculture*, 17 Aug. 2009, https://deepgreenpermaculture.com/2009/08/17/companion-planting/#:~:text=Companion%20planting%20is%20the%20practice.

"Choosing the Right Location for Your Vegetable Garden." *Newsroom*, 7 Apr. 2020, https://sebsnjaesnews.rutgers.edu/2020/04/choosing-the-right-location-for-your-vegetable-garden/

"Companion Planting Chart for Vegetable Garden: Tomatoes, Potatoes, and More! | Guide to Companion Planting | the Old Farmer's Almanac." *Www.almanac.com*, www.almanac.com/companion-planting-guide-vegetables#:~:text=The%20Companion%20Planting%20Chart%20lists.

Hailey, Logan. "15 Companion Planting Mistakes to Avoid This Season." *All about Gardening*, 22 June 2022, www.allaboutgardening.com/companion-planting-mistakes/.

"History of Companion Planting – How Did Companion Planting Start." *Gardening Know How*, 22 Mar. 2022, https://blog.gardeningknowhow.com/tbt/history-of-companion-planting/

"How to Mix Organic Fertilizer with Soil – Foliar Garden." *Foliargarden.com*, https://foliargarden.com/how-to-mix-organic-fertilizer-with-soil/

"How to Use Cover Crops to Improve Soil." *FineGardening*, 23 Sept. 2020, www.finegardening.com/project-guides/gardening-basics/how-to-use-cover-crops-to-improve-soil.

https://www.facebook.com/marthastewart. "The Difference between Deadheading and Pruning – and How to Use Each for Healthier Plants and Flowers." *Martha Stewart*, www.marthastewart.com/8041967/deadheading-pruning-differences.

https://www.facebook.com/thespruceofficial. "Companion Plants Repel Garden Pests and Attract Beneficial Insects." *The Spruce*, 2019, www.thespruce.com/companion-planting-1402735.

https://www.facebook.com/WebMD. "Benefits of Companion Planting." *WebMD*, www.webmd.com/a-to-z-guides/benefits-of-companion-planting#:~:text=One%20of%20the%20few%20companion.

https://www.howstuffworks.com/hsw-contact.htm. "HowStuffWorks Answers Your Gardening Questions." *HowStuffWorks*, 21 Aug. 2007, home.howstuffworks.com/gardening/garden-design/gardening-questions-answered.htm.

Judd, Angela. "Garden Troubleshooting Guide: How to Identify & Solve Common Garden Problems." *Growing in the Garden*, 7 Jan. 2022, https://growinginthegarden.com/garden-troubleshooting-guide-how-to-identify-solve-common-garden-problems/

margo. "The Complete 2023 Organic Fertilizer's Guide for Plants." *HomeBiogas*, 18 Jan. 2023, www.homebiogas.com/blog/organic-fertilizer-for-plants/.

"Organic Fertilizer vs. Chemical Fertilizer | Kellogg Garden Organics™." *Kellogggarden.com*, https://kellogggarden.com/blog/fertilizer/the-advantages-of-organic-fertilizers-over-chemical-fertilizers/

Poindexter, Jennifer. "10 Tips to Harvest Your Garden Vegetables Perfectly and on Time." *MorningChores*, 4 Apr. 2018, https://morningchores.com/harvesting-your-garden/

"Should You Plant Seeds or Plants in Your Garden? • Gardenary." *Gardenary*, www.gardenary.com/blog/should-you-plant-seeds-or-plants-in-your-garden.

"Soil Preparation: How Do You Prepare Garden Soil for Planting?" *Almanac.com*, www.almanac.com/soil-preparation-how-do-you-prepare-garden-soil-planting.

Walliser, Jessica. "Plant Covers to Protect the Garden from Pests and Weather." *Savvy Gardening*, 29 Apr. 2022, https://savvygardening.com/plant-covers/

When and How to Water Your Seedlings and Seeds the Right Way. 13 Feb. 2023, www.gardeningchores.com/watering-seedlings/.